AF615634

Principles of Reinforced Concrete Design

Mete A. Sozen
Toshikatsu Ichinose
Santiago Pujol

CRC Press is an imprint of the
Taylor & Francis Group, an **informa** business

CRC Press
Taylor & Francis Group
6000 Broken Sound Parkway NW, Suite 300
Boca Raton, FL 33487-2742

Printed on acid-free paper
Version Date: 20140319

International Standard Book Number-13: 978-1-4822-3148-9 (Hardback)

Library of Congress Cataloging-in-Publication Data

Sozen, Mete Avni, 1930-
Principles of reinforced concrete design / authors, Mete A. Sozen, Toshikatsu Ichinose, Santiago Pujol.
pages cm
Summary: "The book covers fundamental concepts related to mechanics and direct observation, and those required to design reinforced concrete (RC) structures. Codes change constantly over time depending on factors that have little to do with the fundamental concepts mentioned, and have more to do with the markets, construction practices, and transient academic views. For beginning engineers it is difficult to distinguish between rules based on consensus (codes) and fundamentals. This book focuses on the latter to prepare the engineer to use and adapt to the constant changes of the former. "-- Provided by publisher.
Includes bibliographical references and index.
ISBN 978-1-4822-3148-9 (hardback)
1. Composite construction. 2. Reinforced concrete. I. Ichinose, Toshikatsu. II. Pujol, Santiago. III. Title.

TA664.S67 2014
721'.0445--dc23 2014008075

Visit the Taylor & Francis Web site at
http://www.taylorandfrancis.com

and the CRC Press Web site at
http://www.crcpress.com

Printed and bound in the United States of America by Edwards Brothers Malloy on sustainably sourced paper.

Contents

Preface

We planned and wrote this book to provide essential information needed to proportion reinforced concrete structures subjected to demands that can be expressed in terms of static forces. Two concerns have bounded the material: (1) to provide the needed information within the scope of a four-month semester in a series of sections, most of which will require a single lecture, and (2) to include a minimum of references to design criteria contained in building codes.

We fully appreciate that engineering design is not limited to the realm of science. Because we build primarily from experience (Cross and Morgan, 1932), as members of a profession we are constrained to keep our choices in design within bounds defined by professional consensus and documented in building codes. In the text, we remind the reader to consult the local building code before committing to sizes, strengths, and details. Nevertheless, we do not want the reader to confuse principles based on mechanics, confirmed experience, and experiment with rules based on professional consensus alone.

From our perspective, structural design (choice of size, strength, and framing type) is based on the following:

1. Sense of proportion
2. Understanding of the conditions of static equilibrium and plane geometry
3. Accumulated knowledge on the composite action of concrete and reinforcement derived from mechanics, experiment, and experience

We expect the reader to be proficient in (1) and (2). Unquestionably, success in engineering depends first and foremost on a sense of proportion. That sense is difficult to acquire unless the right questions are asked at the right time, and unless the engineer is familiar with the relevant frames of reference. The engineer needs to have depth in (2) in order to develop confidence and improve his or her choices in (1). This book focuses primarily on issues related to (3).

In conclusion, we thank the reader for his or her interest and acknowledge our debts to our respective universities: Nagoya Institute of Technology, Nagoya, Japan, and Purdue University, West Lafayette, Indiana, as well as to Dr. Cemalettin Donmez of the Izmir Institute of Technology, Urla, Turkey, whose critical comments helped improve the manuscript.

Notation

AREAS, MOMENTS OF AREAS, SECTION MODULUS

A_c	Cross-sectional area of the column core confined by spiral reinforcement
A_g	Gross cross-sectional area
A_s	Cross-sectional area of reinforcing bars (for columns, total area; for beams, area of reinforcement in tension)
A'_s	Cross-sectional area of reinforcing bars in compression
A_{sh}	Cross-sectional area of hoop bar
A_{tr}	Area of transverse reinforcement
A_w	Cross-sectional area of web (or transverse) reinforcement
I_{cr}	Moment of inertia of cracked cross section
I_g	Moment of inertia of gross cross section
S	Section modulus

DIMENSIONS, DEFLECTION, ROTATION

b	Cross-sectional width of a rectangular section
b_f	Flange width of an I- or T-section
b_w	Web thickness of an I- or T-section
c	Distance to neutral axis
c_b	Concrete cover
D	Diameter of test cylinder
d	Effective depth
d′	Depth to compression reinforcement
d_b	Bar diameter
d_c	Diameter of concrete core
H	Height
h	Cross-sectional depth
h_c	Test cylinder height
L	Span
L_d	Development or anchorage length
L_n	Clear span
s	Spacing of hoops or stirrups
t_f	Flange thickness
w_m	Mean crack width
y	Distance to cross-sectional centroid
Δ	Deflection
θ	Slope

FORCES, LOADS, MOMENTS

C Compression force
M Bending moment
M_{cr} Moment at cracking
M_n Nominal capacity to resist bending moment
M_u Bending moment demand
M_y Moment at yield of longitudinal reinforcement
P Concentrated force
P_n Nominal capacity to resist compressive axial load
T Tension force
V Shear force
V_n Nominal capacity to resist shear force
w Distributed load

NONDIMENSIONAL RATIOS

j Ratio of the internal lever arm to the effective depth
k Ratio of the depth to the neutral axis to the effective depth
k_1 Ratio of the area within the compressive stress distribution to area of an enclosing rectangle
k_2 Ratio of the distance between the depth to the line of action of the compressive force and neutral axis depth, assumed to be $k_1/2$
k_3 Ratio of the maximum compressive stress in the compression block to compressive strength determined from 6×12 in. test cylinders
k_b Ratio of the neutral axis depth to the effective depth at simultaneous development of useful limiting compressive strain in the concrete and yield strain in the tensile reinforcement
k_u Ratio of the depth to the neutral axis to the effective depth at the limiting condition
m Creep coefficient
n E_s/E_c
r Transverse reinforcement ratio
w_v/c_v Ratio of the volume of water to the volume of cement
Φ Strength reduction factor
γ_c d'/d
ρ Longitudinal reinforcement ratio (A_s/A_g for columns; $A_s/(bd)$ for beams)
ρ' Compression reinforcement ratio
ρ_b Reinforcement ratio associated with the simultaneous development of useful limiting compressive strain in the concrete and yield strain in the tensile reinforcement
ρ_h Volumetric ratio of spiral reinforcement
ρ_{max} Maximum ratio of longitudinal reinforcement

ρ_{min}	Minimum ratio of longitudinal reinforcement
ρ_t	Ratio of the creep or shrinkage strain at a given time to the total creep or shrinkage strain

UNIT CURVATURES

ϕ	Unit curvature
ϕ_{cr}	Unit curvature at cracking
ϕ_{creep}	Increase in unit curvature caused by creep
ϕ_{inst}	Instantaneous unit curvature caused by sustained loads
ϕ_{shrink}	Increase in unit curvature caused by shrinkage
ϕ_u	Unit curvature associated with limiting strain in the concrete ε_{cu}
ϕ_y	Unit curvature at yielding of longitudinal reinforcement

UNIT STRAIN

ε_c	Maximum unit strain in concrete
ε_{cr}	Unit strain at cracking
ε_{cu}	Limiting unit strain in concrete
ε_{inst}	Instantaneous unit strain in concrete
ε_o	Unit strain at peak stress in concrete
ε_s	Unit strain in longitudinal reinforcement
ε'_s	Unit strain in compression reinforcement
ε_{sh}	Unit shrinkage strain
ε_{su}	Unit strain in longitudinal reinforcement at limiting condition
ε_u	Limiting strain of reinforcement
ε_y	Unit yield strain of reinforcement

UNIT STRESSES, MODULI, UNIT WEIGHT

E_c	Young's modulus of concrete
E_s	Young's modulus of steel
f''_c	Peak stress in unconfined concrete
f'_c	Compressive strength of standard 6 × 12 in. cylinders
f_1	Unit axial stress in confined concrete
f_2	Lateral unit stress in confined concrete
f_c	Maximum compressive unit stress in concrete
f_{ct}	Effective tensile strength of concrete
f_r	Modulus of rupture
f_s	Unit stress in longitudinal steel
f'_s	Unit stress in compression reinforcement
f_{sw}	Unit stress in web or transverse reinforcement
f_t	Tensile strength of concrete

f_{ts}	Splitting strength of concrete
f_y	Unit yield stress of reinforcement
f_{yh}	Unit yield stress of spiral reinforcement
f_{yw}	Unit yield stress of web or transverse reinforcement
μ	Bond stress
μ_a	Allowable bond stress
μ_f	Flexural bond stress
v	Unit shear stress
v_c	Unit shear strength attributed to concrete
v_n	Unit shear strength
v_s	Unit shear strength attributed to transverse reinforcement
w_c	Unit weight of concrete

About the Authors

Mete A. Sozen, S.E. (IL), a graduate of Bogazici University (Istanbul) and the University of Illinois at Urbana–Champaign, is the Kettelhut Professor in the Department of Civil Engineering at Purdue University, West Lafayette, Indiana. He teaches courses on reinforced concrete and earthquake-resistant design. He is a member of the U.S. National Academy of Engineering and the Royal Swedish Academy of Engineering Sciences. He is an honorary member of the Turkish Society for Engineers and Scientists, the American Society of Civil Engineers, the American Concrete Institute, the Architectural Institute of Japan, and the International Association for Earthquake Engineering. Dr. Sozen has been granted honorary degrees by Bogazici University (Turkey), Janus Pannonius University (Hungary), and the Tbilisi Technical University (Georgia). He was included in the Applied Technology Council and the *Engineering News-Record* lists of the top ten seismic engineers of the 20th century.

Dr. Sozen has worked as a consultant with the Veterans Administration, the Department of State, Bechtel, Brookhaven National Laboratory, Consumers Power Co., Electric Power Research Institute, ERICO, Lawrence Livermore Laboratories, Los Alamos National Laboratory, Nuclear Regulatory Commission, SANDIA National Laboratories, U.S. Army Engineering R & D Center in Vicksburg, WJE Engineers, and Westinghouse Savannah River Site. He has worked on design and evaluation of concrete dams with the U.S. Bureau of Reclamation, Pacific Gas and Electric Co., and Southern California Edison.

Toshikatsu Ichinose completed his undergraduate education in architectural engineering at Nagoya Institute of Technology and earned graduate degrees at the University of Tokyo. In 1982, he returned to Nagoya Institute of Technology, where he now teaches structural mechanics and reinforced concrete. Dr. Ichinose has been chairing the committee in charge of RC building code of the Architectural Institute of Japan since 2004. He is proud to be a coauthor of *Understanding Structures* (CRC Press, 2008).

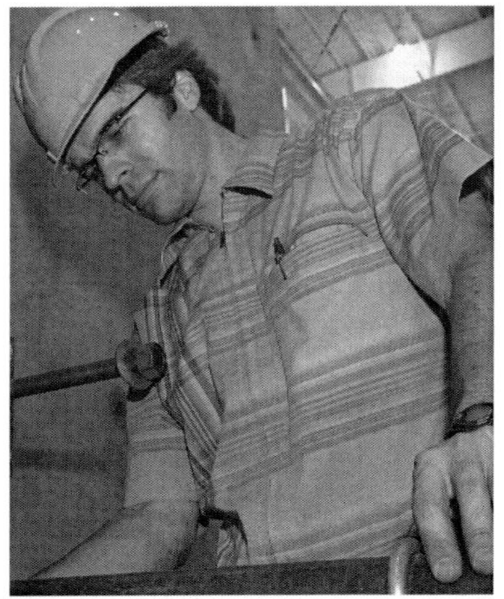

Santiago Pujol is from Medellín, Colombia, where he was first exposed to the reinforced concrete industry by his father, an architect who designed and directed the construction of a number of residential and commercial buildings in the city. He completed his undergraduate education in civil engineering at the School of Mines at the National University of Colombia in 1996. At the time, the School of Mines had a five-year undergraduate program designed after European academic models. Damage caused by earthquakes in Colombia motivated him to earn an MS and PhD from Purdue University, West Lafayette, Indiana. Dr. Pujol worked with a forensic firm in San Francisco, California, from 2002 to 2005. During that time, he traveled to Japan and started lasting collaborations with Japanese engineers whose wisdom and attention to proportions and aesthetics has a strong influence on his views. In 2005, he returned to Purdue University, where he currently teaches reinforced concrete and experimental methods and conducts research dealing with the effects of blasts and earthquakes on buildings, and basic questions on the mechanics of concrete structures.

1 A Brief History of Reinforced Concrete

Cement, in its various forms, has a surprisingly long history that spans back to prehistoric times. In contrast, the development of reinforced concrete is relatively recent. It is interesting to note that manure, clay, stone, timber, and metal were used "as is" in building. It took human creativity to put metal and concrete together to give it special properties and to invent or reinvent ways of using it to serve society.

Invention in reinforced concrete started in the mid-nineteenth century with Joseph Lambot's boat (Figure 1.1) and continues to our day. The brief view of the history of reinforced concrete in this chapter is intended to emphasize to the student that there is still room for creativity in uses of reinforced concrete.

Lambot's main interest was agriculture. That was why his first commercial product in reinforced concrete was a container for oranges. The next notable inventor in reinforced concrete was another agriculturist, Joseph Monier, who also started experimenting with reinforced concrete containers in 1867. In 1868, he projected his concept to pipes. After having tried precast panels for architectural facades in 1872, in 1873 he expanded his container concept to build a large silo to hold cement. His success with the silo inspired him to build a bridge in 1875 at the Château de Chazelet, France. The bridge led him to patent the concept of the reinforced concrete beam. His contribution was a series of pragmatic inventions that appeared to be the product of cut-and-try thinking.

Francois Hennebique, inspired by Monier's successes, developed a scientific approach to the proportioning of reinforced concrete elements to obtain a patent in 1879, one that was denied to him in favor of Monier seven years later. The Hennebique construction company helped make reinforced concrete a serious construction material in Europe. Hennebique's signal triumph was the initiation of a technical journal on reinforced concrete in 1896.

Hennebique's attempts at providing an intellectual underpinning for design attracted German engineers to consider reinforced concrete. Ritter (1899) recognized the nonlinearity of concrete at higher stresses and was the first to visualize a reinforced concrete beam not as it is, but as a truss. His ideas were further developed by Moersch (1902).

The initial attraction for using reinforced concrete in building construction may have been its fire resistance, but its increasing popularity was due to the creativity of engineers who kept extending its limits of application. The French engineer Francois Coignet expanded the use of reinforced concrete in European buildings. In the United States, William E. Ward built the first landmark building structure in reinforced concrete in Port Chester, New York (Figure 1.2). It did not take very long

FIGURE 1.1 Boat built by Joseph Lambot, 1849. (Photo courtesy of the Musée du pays Brignolais.)

for enterprising engineers in the United States to push the limit to 15 stories or 210 ft (Ingalls Building, Figure 1.3). In the same year, C. A. P. Turner built his first flat slab that popularized the construction of reinforced concrete buildings throughout the world because of its phenomenal economy and convenience (Soven and Siess, 1963). The next important height achievement was in Chicago's Executive House Hotel that reached 371 ft, completed in 1958. Convenient access to 6000 psi concrete and the concept of the flat plate (derived from the flat slab) enabled William Schmidt to conceive the proportions of the Lake Point Tower (645 ft) in 1964, which was noteworthy not only because of its height, but also because of its low cost (Figure 1.4).

The height of Lake Point Tower was topped by the White Castle, also in Chicago, which reached 961 ft. The next notable jumps were to 1476 ft in the Petronas Twin Towers in Kuala Lumpur, Malaysia, and 2684 ft in the Burj Khalifa building in United Arab Emirates. Building height, also influenced by the state of the economy, is a good, if incomplete, indicator of the advances in building technology (Figure 1.5). Reinforced concrete became a successful building medium because of continual improvements in its strength, its economy, its durability, its fire resistance, and its beauty (Figure 1.6). The Monier and Hennebique experiments with bridges in reinforced concrete introduced the other aspect of reinforced concrete: its "plasticity" in encouraging elegance in the hands of engineering artists such as Robert Maillart of Switzerland (Figure 1.7) and Eduardo Torroja of Spain (Figure 1.8).

In the United States, creativity of the builders in reinforced concrete was supported by research in U.S. laboratories. As opposed to conceptual developments in Europe influenced strongly by the theory of elasticity, A. N. Talbot developed design methods based primarily on observation.* The effect of the water–cement ratio on

* His influence lasted in the United States until the end of the twentieth century, when research became an academic requirement. One of his students, Mikishi Abe, took Talbot's message of pragmatism to Japan, where it is still strong.

FIGURE 1.2 First reinforced concrete building in the United States. (Port Chester, New York, 1875). (Photo courtesy of Daniel Case/Wikimedia Commons.)

concrete strength was discovered by Duff Abrams in the early twentieth century. In a project sponsored by the membership of the American Concrete Institute, the foundations for consistent design of reinforced concrete columns were developed by F. E. Richart, who made durable contributions to many aspects of design.

Hardy Cross worked on arches and frames to invent methods that enabled engineers to deal with continuity, an essential requirement for designing reinforced concrete structures. Cross's creation of the moment-distribution method popularized the use of reinforced concrete structures. However, he is reported to have said that his method was too exact for inexact structures. He thought that knowing the conditions of equilibrium and developing a sense of deflected shapes of structural elements would suffice for proportioning of continuous frames. All young engineers would benefit immensely from reading Chapter 2 of the Cross–Morgan opus on continuous frames (Cross and Morgan, 1932). Westergaard was successful in simplifying plate theory for use in design. N. M. Newmark and C. P. Siess took the lead in developing simple design methods for reinforced concrete structures subjected to dynamic loads such as earthquake and blast and for prestressed concrete buildings and bridges.

Creativity in reinforced concrete was also observed in the construction arena. Slip-form construction was first used to build silos. Its later adaptation to build

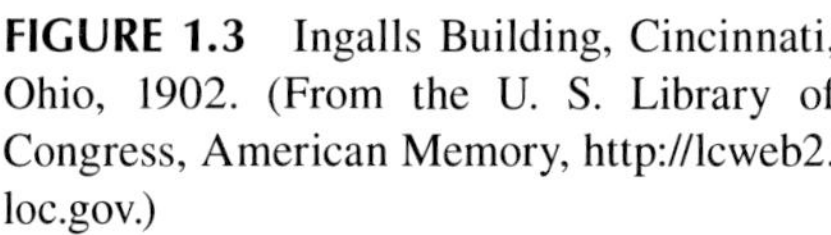

FIGURE 1.3 Ingalls Building, Cincinnati, Ohio, 1902. (From the U. S. Library of Congress, American Memory, http://lcweb2.loc.gov.)

FIGURE 1.4 Lake Point Tower, Chicago, Illinois, 1964. (From John Kershner/Shutterstock.com.)

reinforced concrete walls revolutionized the construction scene. This was matched by developments in precast construction.

Because concrete has a tendency to change its volume with time, methods had to be developed to control negative effects of such changes, especially in massive concrete structures such as dams. The Bureau of Reclamation in Denver, Colorado, provided the intellectual underpinnings for the achievement of such magnificent projects as the successful construction of Boulder and Grand Coulee Dams in the 1930s.

Reinforced concrete continues to be improved. During the last decade of the twentieth century, the compressive strength of concrete moved well above 10,000 psi from a typical 4000 psi. Self-consolidating concrete made casting of complicated shapes an easy task. Every aspect of the short history of reinforced concrete indicates that the inventions will continue.

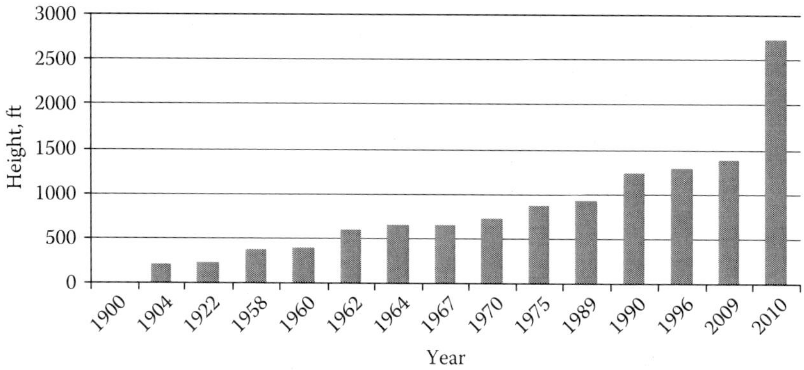

FIGURE 1.5 Changes in maximum height of reinforced concrete buildings.

FIGURE 1.6 Nariwa Museum designed by Tadao Ando, Japan. (Photo courtesy of N. Hanai.)

FIGURE 1.7 Bridge designed by Maillart, Switzerland. (From Rama [CC-BY-SA-2.0-fr (http://creativecommons.org/licenses/by-sa/2.0/fr/deed.en)], Wikimedia Commons.)

FIGURE 1.8 Torroja's Aqueduct in Alloz, Spain. (Photo courtesy of Juan Manuel Galindo de Pablos.)

BARE ESSENTIALS

Reinforced concrete became a successful building medium because of continual improvements in its strength, its economy, its durability, its fire resistance, and its beauty.

The ongoing momentum of developments in its strength, its durability, and its pliability to fit creative architectural demands suggests that this structural composite continues to be wide open to invention.

2 Structural Framing in Reinforced Concrete

One of the attractions of reinforced concrete is its plasticity: It can be built conveniently in any shape. A reinforced concrete structure can be assembled using any combination of planar (shells, walls, slabs) and linear (columns, girders) elements. It offers a multiplicity of options to creative architects (Figure 2.1) and engineers (Figures 1.6 and 1.7).

Reinforced concrete offers options in economy too. The simplest reinforced concrete building structure, in terms of geometry, is one composed of footings, columns, and flat plates. As an assembly invented by C. A. P. Turner, this type of structure is called a flat plate (Figure 2.2) or a flat slab if the columns have capitals and the slabs have increased depth around the capitals (Figure 2.3). The flat plate and the flat slab are economical because they require smaller investment in formwork than do other structural configurations in reinforced concrete, because they permit clean ceilings, and because structural elements do not interfere with conduits.

The most critical component of the flat-plate structure is the connection between the slab and the columns. This connection can be improved with the use of capitals and drop panels. (Initially, the thickened region around the column was achieved by dropping the form panel. Hence, the thickened region is currently called a drop panel; Figure 2.3.) Capitals and drop panels are often used in parking and industrial structures requiring larger spans and heavier loads.

As an alternative to the use of capitals and drop panels, and to increase the stiffness and strength of the flooring system, a series of parallel beams can be cast to support the slabs (Figure 2.4).

A combination of beams and joists can also be used to produce strong and stiff floors if the increases in the costs of formwork and labor are not prohibitive (Figure 2.5).

Another option to stiffen and strengthen a slab is to support it with beams running along column lines in two perpendicular directions (beams identified by dark and light colors in Figure 2.6). These beams are connected to columns to form frames that add lateral strength and stiffness to the structure.

In this system, load acting on a slab panel, an area defined by bounding column centerlines on four sides, is resisted by beams running in two directions (dark and light beams in Figure 2.6). This type of slab is called a *two-way* slab in reference to the fact that load is assumed to be transmitted from the slab to the beams in two directions (Figure 2.7a). In slabs supported by beams parallel to one another, the load is transferred from the slab to beams in a single direction (Figure 2.7b). Slabs with

FIGURE 2.1 A reinforced concrete structure in Japan designed by Kotaro Ide/ARTechnic architects. (Photograph courtesy of Nacasa & Partners Inc.)

FIGURE 2.2 Flat-plate structure.

beams running in a single direction, whether they have joists or not, are therefore called *one-way* slabs.*

The strength and lateral stiffness provided by frames are necessary to resist wind, blast, soil, and earthquake demands. Depending on the expected magnitude of lateral loads, the addition of structural walls, also called shear walls (Figure 2.8), may be necessary.

Both vertical loads, such as self weight and weight of building contents and occupants, and lateral loads, such as loads generated by wind or earthquake, need to be transferred to the foundation. Foundations may consist of a single reinforced concrete mat or a series of footings. Footings may or may not be connected by grade beams (Figure 2.9), depending on soil conditions.

Arriving at an effective framing concept for a structure requires selection of the appropriate structural system having the right proportions. It is wise to have an idea

* It is of interest to note that historical development followed a different course. Initially, structural framing was based on beam-and-post combinations in timber. Introduction of iron and steel emphasized the use of one-way construction where the load was carried in one direction at a time, from the floor planks to the girders, from the girders to the columns. The first experiments in reinforced concrete used the same system. The two-way slab followed. The flat plate, construction without girders, was invented much later.

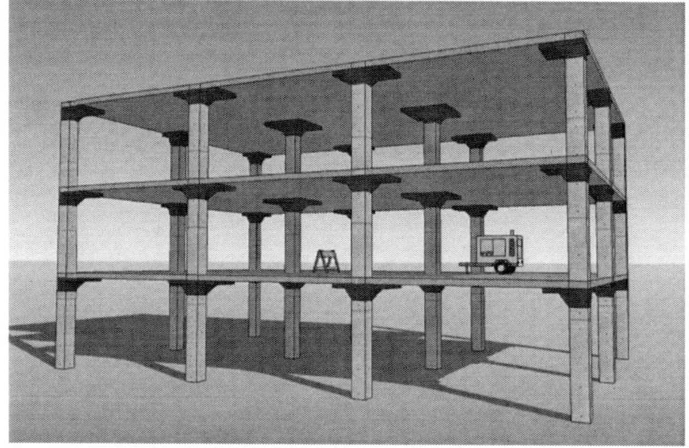

FIGURE 2.3 Column capitals (in dark color).

FIGURE 2.4 Reinforced concrete structure with beams (in darker color) in one direction.

FIGURE 2.5 Structure with beams (in dark color) and joists (in light color).

FIGURE 2.6 Structure with two-way slabs.

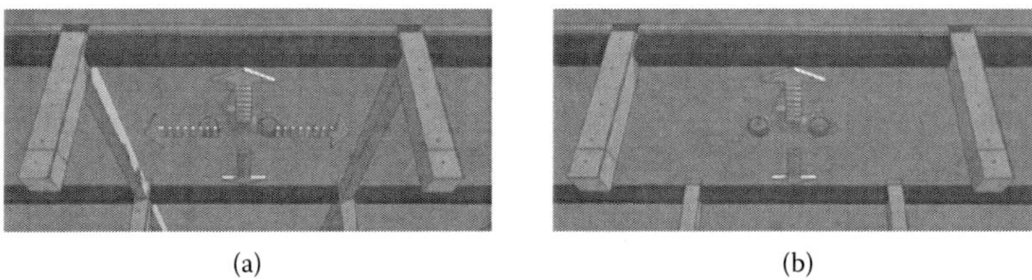

(a) (b)

FIGURE 2.7 (a) Two-way slab. (b) One-way slab.

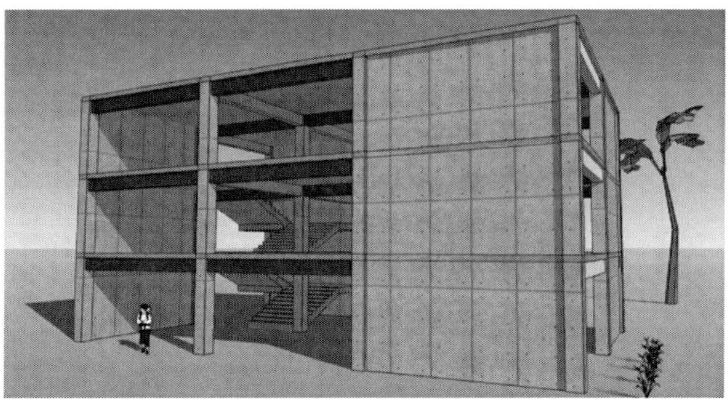

FIGURE 2.8 Structure with shear walls.

of the right proportions of the structural elements before undertaking calculations for design. How do we develop an idea of the right proportions? Simply by looking at successful structures and observing their proportions.

The following are *guidelines* for selection of preliminary proportions. They are not absolute rules. The proportions of building structures vary from place to place,

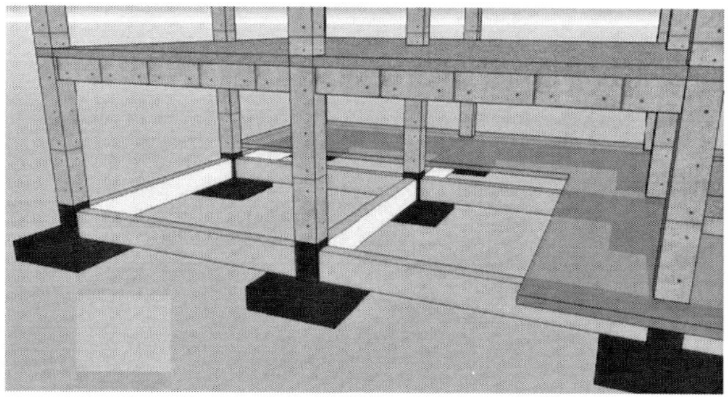

FIGURE 2.9 Foundation.

depending on local demands and experience. It is good to realize that physics is universal but engineering is local. Existing structures are the best source of information on proper proportions.

1. The depth of a typical continuous reinforced concrete beam is L/12. (In the Imperial Unit System, the depth in inches should be approximately the same as the span in feet, h in. ~ L ft.)
   (L = span length)
2. The width of a reinforced concrete beam is ~h/2.
   (h = beam depth)
3. The clear height of a reinforced concrete column (distance from top of slab to bottom of beam or slab above) is four to seven times the depth of its cross section.
4. The thickness of a two-way concrete slab ranges from L/30 to L/36.
   (L = span length in inches)
5. The thickness of a one-way concrete slab without joists is L/20 to L/24.
   (L = span length in inches)
6. The unit weight of a reinforced concrete building is 150 to 200 psf.

BARE ESSENTIALS

Existing structures are the best source of information on proper proportions.

Proper proportions may change from region to region, depending on culture and history, as well as sensitivity to loading demands such as fire, wind, and earthquake.

EXERCISES

1. Select preliminary dimensions for the columns of a 4-story frame with four 25 ft center-to-center spans. Assume that the *clear* story height is required to be 12 ft.
2. Select preliminary dimensions for the beams of the frame described in (1). Draw the resulting frame and element cross sections to scale.
3. Visit five cast-in-place reinforced concrete buildings. Record the ratios of measured or estimated column clear height to cross-sectional depth in different stories and, if possible, beam or slab span to thickness. Indicate in your record the use of the structure and the location of the element (e.g., interior column in first story). Try to focus on elements of the main structural framing system: exclude stairs, bridges, walkways, and canopies.

3 The Design Process

The initial phase in proportioning a structure, modestly called preliminary design, is the most important one in the conception of a structure. Preliminary design, mostly an art, exerts a strong influence on the final product that results from generally analytical evaluations of the initially chosen framing type and section dimensions. Perhaps the best witness to the relative importance of preliminary design is contained in the engineering adage that if a calculation produces a result that differs by more than 15% from the engineer's initial guess, then there is likely to be something inadequate in the engineer's judgment, in the calculation, or in both. A careful check or consultation with an experienced engineer is in order.

Design goals were stated succinctly by Vitruvius in 1 BC: A structure should have *firmitas* (strength), *utilitas* (utility or serviceability as well as sustainability and resilience), and *venustas* (beauty). Vitruvius's goals continue to be relevant today and deserve discussion.

Strength refers to the ability of the structure to resist loads. Throughout this book, we use the word *load* to refer to forces that need to be resisted by the structure. Structures may be subjected to force, displacement, and heat energy demands. Examples of displacement demands are differential settlement of footings (Figure 3.1), volumetric changes of materials, and earthquake effects. In fact, the load demand is better generalized in terms of energy demand (product of force and displacement) as well as time. In this book, we focus on static loads applied slowly and monotonically (without repetitions or reversals). Fire is a threat more frequently encountered than overload.

Self weight and the weight of finishes, partitions, and permanent equipment are examples of sustained load. Sustained loads are referred to as *dead loads* (DLs). Other weights that require consideration include the weights of occupants, vehicles, furniture, and portable equipment. The location and duration of these loads may change, and consequently, they are called *live loads* (LLs). Because of their sources, uncertainties related to live loads exceed uncertainties related to dead loads. Other demands on structures include wind, flood, earthquake, snow, ice, fire, and pressures caused by soil and fluids.

Serviceability covers the functionality, durability, and sustainability of the structure. The design of a reinforced concrete structure should ensure that (1) deflections and vibrations do not create discomfort, (2) cracks—if they occur—remain tolerable, and (3) reinforcement corrosion is slowed down. All of these requirements refer to daily use of the structure. Strength requirements, on the other hand, refer to events, with a very low probability of occurrence, that can demand the full capacity and tenacity of the structure.

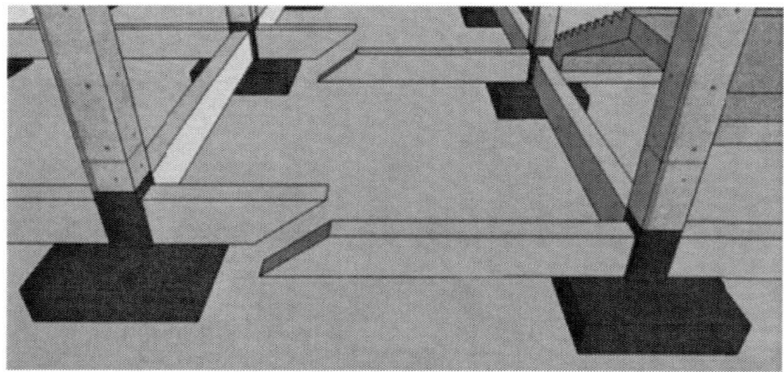

FIGURE 3.1 Idealized failure of grade beams caused by differential settlement.

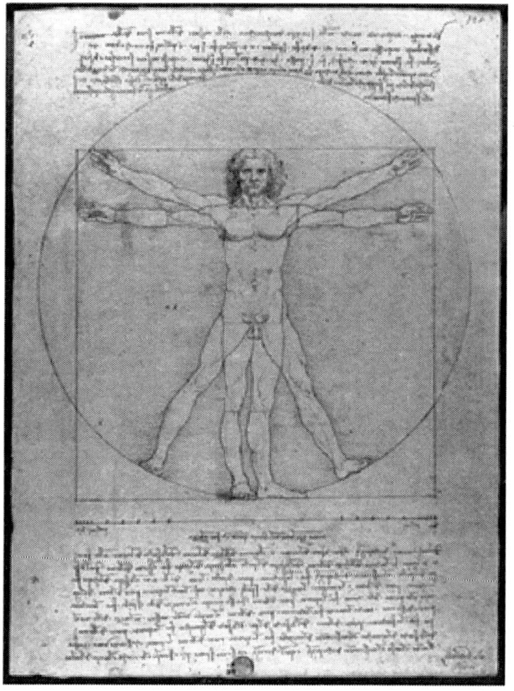

FIGURE 3.2 Da Vinci's illustration of proportions described by Vitruvius.

Beauty: The attention that Vitruvius gave to proportion is evidenced by his studies of the proportions of the human body illustrated later by Leonardo Da Vinci (Figure 3.2). A good design should produce a structure that not only works but also looks right. This requirement can seldom be satisfied through calculation alone. The student should realize that, here, design departs from analysis. Analysis appears to be exact and produces usually a single answer (to our idealized versions of real problems). Design is subjective, and in this and other respects, it is closer to art than to science.

The design process includes five steps:

1. Definition of the use of the structure and selection of design demands
2. Selection of framing types and preliminary dimensions
3. Analysis
4. Selection of reinforcement and final dimensions
5. Detailing

Every one of these steps is crucial and requires judgment and experience. All steps should be geared toward producing a structure that is strong, useful, and beautiful. Because the degree to which these goals are achieved may affect the safety and interests of the public and because design extends beyond science, a set of minimum requirements are summarized in documents that are meant to express the consensus of the profession and are enacted into law by local authorities (building codes). In using such documents the designer must bear in mind that these requirements are meant to be the minimum. The goal of good design is more than simply satisfying the code requirements.

3.1 DEFINITION OF THE USE OF THE STRUCTURE AND SELECTION OF DESIGN LOADS

The intended use of the structure determines the design loads. For instance, a warehouse for storage of cement is very likely to experience much larger loads than a residential building. Building codes regulate the magnitude of the loads to be considered in design. Table 3.1 shows examples of service level live loads used in design.

The unit loads listed in Table 3.1 do not represent realistic estimates or even best estimates of the expected load. Rather, they refer to reasonable upper bounds. The current concept of design for strength involves the rather vague idea of setting the intersection of the highest conceivable demand with the lowest credible strength to a tolerably low probability. To achieve this ideal limit, the specified loads are boosted by load factors, and the calculated strengths are reduced by strength reduction factors. Experience has played a dominant role in the determination of these factors. However, they are often presented in reference to estimated or assumed probabilities of failure. The most frequent cause of damage under normal conditions of loading is an error in design or construction.

TABLE 3.1
Service Level Live Loads

Private residential spaces (other than balconies)	30–40 psf
Parking structures	50 psf
Offices	50 psf
Corridors and balconies (larger than 100 ft^2)	100 psf
Stack rooms in libraries	150 psf

It is relevant to note here that structural engineering is said to have become a specialty within the building profession as a result of the debate that ensued in relation to the restoration of St. Peter's Dome in Rome during the eighteenth century. Three scientists retained by Pope Benedict XIV maintained that the anticipated negative effect of cracks observed in the dome could be eliminated by adding chain reinforcement circumferentially at the base of the dome. The cross-sectional dimensions of the needed chains were determined by a simple application of what we currently call the virtual work theorem. That development was instrumental in initiating a thinking trend that created structural engineering as a profession.

The development in eighteenth-century France of the theory of elasticity had a profound effect on the practice of structural engineering. Design was conducted in the linear elastic range of response using allowable stresses that were determined by dividing the assumed strengths of the materials by what were considered to be factors of safety. This approach, called working stress design, has been successfully used in Japan through the present day. In the United States, the approach was changed in mid-twentieth century to favor what is called strength design. At that juncture, there was a division of thought in design of reinforced concrete: Should the factor of safety relate to the strength of individual sections or to the strengths of the materials constituting the section? Debate within the professional societies resulted in the choice of section strength as a criterion. The reason was approximately as follows.

The demands on and resistances of a reinforced concrete structure are classified with respect to how well they are understood, as shown in Table 3.2.

Examples of well-understood demands are self weight of a floor and pressure on a wall retaining a static fluid. Examples of poorly understood demands are those of natural hazards such as earthquake and wind.

Examples of well-understood resistance sources in reinforced concrete elements are those for axial load and bending. Examples of poorly understood sources of resistance are shear and bond strength.

Thus, cases that fall in intersection 1 in Table 3.2 may be dealt with using relatively low factors of safety, and those that fall in intersection 4 require relatively high factors of safety. Cases that fall in intersections 2 and 3 require intermediate factors of safety. The approach described leads to load factors such as 1.2 for self weight and 1.6 for live load and strength reduction factors such as 0.9 for flexure and 0.75 for shear strength. Reductions in load factors are allowed for cases where loads of different origins are expected to act on the structure simultaneously on the basis of the

TABLE 3.2
Classification of Demands and Resistances

	Demand	
Resistance	**Well Understood**	**Poorly Understood**
Well understood	1	2
Poorly understood	3	4

argument that it is unlikely to have, say, a record snowfall on the same day there is a very strong earthquake.

Load amplification and strength reduction factors in prevailing codes are based on the approach described above, even though some explain it in simple pragmatic terms and some by estimated probabilities. Whatever the approach adopted in the local building code, the licensed structural engineer must design within the limits set by that code.

Values of the overall factor of safety (the ratio of load factor to strength reduction factor) are selected based on factors such as the construction history of a given region, the observed quality of workmanship, and the willingness of the public to take risks. Its ranges depend on regional traditions, even though they may be expressed in terms of probabilities. The choice of the level of the overall factor of safety depends not only on strength but also on issues related to serviceability and sustainability. If we wish to have the structure survive until the end of time and to survive every possible demand, the factor of safety needs to be very high, to the point that the design product may be unbuildable. We need to compromise. That brings up an interesting contradiction. We need to negotiate a factor related to failure, not to safety, but we call it a factor of safety. Strictly, the factor of safety admits a certain likelihood of failure. What is a reasonable level for the overall factor of safety? This question does not have a single answer. For instance, factors of safety have ranged from approximately 2 to 4 for columns and 1.2 to 2 for beams.

The overall factor of safety is, in general, influenced by three factors: (1) estimate of the consequences of failure, (2) estimate of the error magnitude in resistance, and (3) estimate of the error magnitude in demand.

1. Consequences of failure: Design deficiencies in certain members may result in disproportionate effects. For example, if a continuous beam is overloaded at its mid-span in flexure, the consequence is likely to be limited to excessive deflection for that particular beam. The distress is localized and does not necessarily affect other parts of the structure. On the other hand, if a lower-story column fails, it may take out with it not only the girders it supports, but also part of the structure.

   Consequently, the factor of safety assigned to a column should be more than that assigned to a beam.
2. Unpredictability of resistance: Consider a simple hanger made of a single reinforcing bar. The resisting force of the bar is readily determined as the product of the nominal cross-sectional area of the bar and the yield stress of the material. The unpredictability involved would rest primarily in the scatter of the yield stress of the material unless design errors are made in the connections. There would be little error introduced by theory in this case.

   Next consider the shear strength of a highway girder. In this case we have to deal with scatter in the tensile strength of the concrete. We also have to deal with construction errors in the dimensions of the girder and theoretical errors in determination of the shear strength compounded by random loading at different speeds.

The unpredictability of the response of the highway girder would be considerably more than that of the strength of the hanger with adequate connections.

3. Unpredictability of demand: Consider the self weight of a reinforced concrete slab. The demand of the self weight would, unless a serious construction error occurs, not be expected to vary by more than a small fraction of its nominal weight. On the other hand, the same cannot be said of, for example, wind loading. The factor of safety for wind load would be set at a higher level than that for self weight.

3.2 SELECTION OF FRAMING AND INITIAL DIMENSIONS

Selection of initial dimensions requires judgment that derives from experience and study. Examples of guidelines to select initial dimensions were provided in Chapter 2. Initial dimensions are usually selected paying special attention to serviceability limits. Deficiencies in strength are more likely to be discovered during the analysis stage than deficiencies in serviceability.

3.3 ANALYSIS

Once loads, framing, and initial dimensions are selected, static linear* analyses are carried out to estimate the magnitudes of the bending and torsional moments, and shear and axial forces that structural elements should be proportioned to resist.

These analyses should consider the fact that certain arrangements of live loads may be more critical than others. For instance, the load pattern shown in Figure 3.3a causes a maximum positive bending moment M_{A1} equal to the difference between the static moment $wL_n^2/8$ and the end moment M_{B1}. The end moment increases as the restraint against rotation provided by the columns and the contiguous beams is increased. This restraint can be increased by loading the adjacent spans as shown in Figure 3.3b. The magnitude of the end moment in this case M_{B2} is greater than the magnitude of the end moment M_{B1} in the previous case. Because equilibrium requires the static moment $wL_n^2/8$ to remain constant, we conclude that the maximum positive moment for case b is smaller than the maximum moment for case a ($M_{A1} > M_{A2}$).

It is the responsibility of the designer to consider the loading pattern resulting in the highest demand at each critical section (sections near supports and sections near mid-span). The designer should therefore work with envelopes of moment and shear diagrams such as those shown in Figure 3.3c and d. These figures show upper-bound estimates of peak moments and peak shear forces. These moments and shear forces have been estimated† considering different arrangements of live loads distributed uniformly along each span. The estimates in Figure 3.3 are applicable to frames with three or more bays. The lengths of contiguous bays should not differ by more than 20%. Negative moments should be computed using the average length of contiguous spans.

---

* See Chapter 22 for how to consider nonlinearity in the analysis and design of continuous beams.

† *Building Regulations for Reinforced Concrete*, American Concrete Institute Committee 318, Detroit, Michigan, 1941.

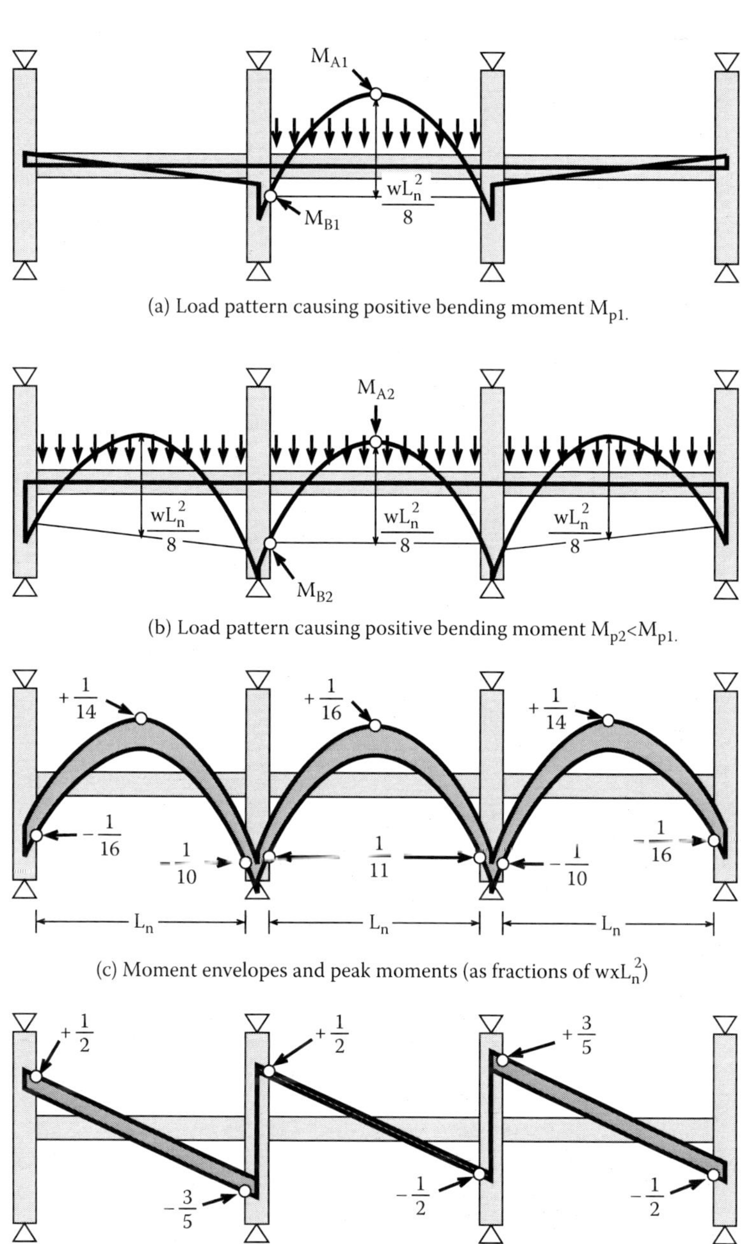

(a) Load pattern causing positive bending moment M_{p1}.

(b) Load pattern causing positive bending moment $M_{p2} < M_{p1}$.

(c) Moment envelopes and peak moments (as fractions of wxL_n^2)

(d) Shear envelopes and peak shear forces (as fractions of wxL_n)

FIGURE 3.3 Approximate moments and shears for design of continuous beams and one-way slabs (L_n = clear span).

3.4 SELECTION OF REINFORCEMENT AND FINAL DIMENSIONS

The amount of reinforcement is chosen in relation to the required strength. The following equation is the basis of design for flexural strength:

$$M_u \leq \Phi M_n \tag{3.1}$$

It states that demand (M_u)—computed for factored loads using linear static analysis—shall not exceed capacity (M_n) modified by a strength reduction factor Φ. This expression refers to bending moment, but the concept applies to shear ($V_u \leq \Phi V_n$), torsion ($T_u \leq \Phi T_n$), and axial load ($P_u \leq \Phi P_n$). Φ is smaller than 1 and reflects uncertainties involved in the estimation of strength and the potential consequences of failure.

In current practice, strength reduction factors, or Φ, range from 0.9 for beams with moderate amounts of reinforcement to 0.6 for structural walls subject to earthquake demands. Columns carrying large axial loads are also assigned low strength reduction factors (0.65 or 0.75, depending on their configuration) because of the potential consequences of their failures. These factors vary over time as academic views and economic conditions change. But the overall factor of safety based on nominal quantities, the ratio of average load factor to average Φ, has remained close to 1.5.

3.5 DETAILING

Detailing refers to selection of reinforcement length, diameter, configuration, and spacing. Detailing is arguably the most common source of errors in reinforced concrete. The subject is treated throughout this text, but there are two basic rules that can be offered here: (1) it is preferable to use more bars of small diameter than fewer bars of large diameter, and (2) reinforcement should be provided across any plane of potential cracking.

EXAMPLE 3.1

For the normal weight reinforced concrete (RC) structure shown (Figure 3.4), compute the design bending moment demand at section A for beam AB. Consider dead loads (DLs) and live loads (LLs). Ignore other loads. Assume the design live load is 50 psf and the superimposed dead load is 20 psf. Use factored loads equal to 1.2 DL plus 1.6 LL. Assume the unit weight of reinforced concrete is 150 pcf.

SOLUTION

The weight of the slab is (7/12) ft × 150 pcf = 88 psf.

The design load *for the slab* is 1.2 DL + 1.6 LL = 1.2 × (88 psf + 20 psf) + 1.6 × (50 psf) = 210 psf (>1.4 DL)

The design load *for the beam,* assuming it carries a 20 ft wide slab strip plus self weight, is w_u = 210 psf × 20 ft + 1.2 (15 × 12 in.) 1 ft²/144 in.² × 150 pcf = ~4400 plf = 4.4 kip-ft

The design moment is (Figure 3.3) $M_u = -w_u \times L_n^2/10$ = −4.4 kip-ft (24 ft − 18/12 ft)²/10 = ~−220 kip-ft

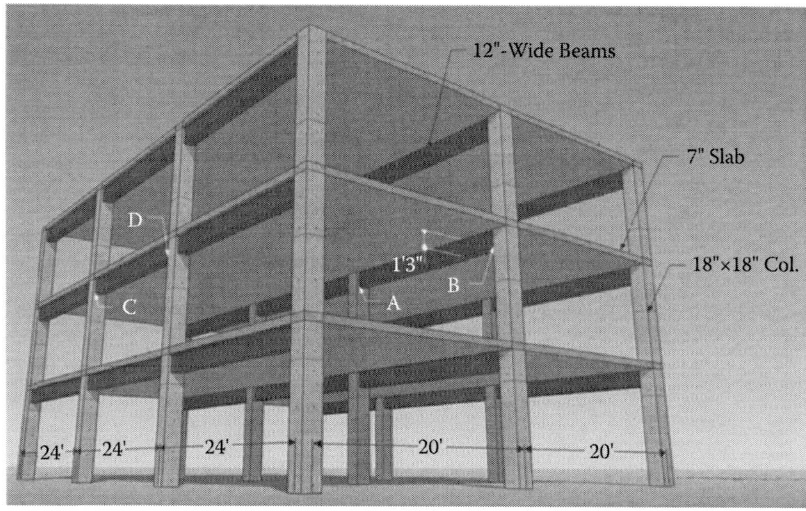

FIGURE 3.4 Example.

> **BARE ESSENTIALS**
>
> Design involves five steps: determination of use, selection of framing and dimensions, analysis, selection of reinforcement, and detailing. Each step is geared toward producing a structure that is strong, serviceable, and beautiful. Strength is mainly understood in terms of load and displacement. Serviceability refers to the need to control deflections and crack widths. Beauty in this context means that the structure needs to look right.

EXERCISES

For the previous example, compute the design bending moments (in kip-ft) at all critical sections for

1. An exterior beam
2. The interior beam
3. A representative 1 ft wide strip of the slab

4 Properties of Steel Reinforcement

Round steel bars with surface deformations (Figure 4.1) are used generally to reinforce concrete in tension, compression, and shear. They are also used to confine concrete. In all cases, the bars are intended to work in axial tension or compression.

Representative mechanical properties of reinforcing bars are shown in Figure 4.2 in reference to a tensile test. The plot shows the changes in unit strain as an axial tensile force is applied on the bar monotonically and slowly. The vertical axis indicates the tensile force divided by the nominal cross-sectional area of the bar. This quantity, which is a unit stress, is usually referred to as stress. It has units of force per unit area. The extension of the bar divided by its original length is referred to as unit strain, or simply as strain. It has no units. Sometimes it is also called engineering strain to distinguish it from true strain, a definition very seldom used in structural engineering practice. True strain is the unit strain obtained as the ratio of deformation increment to the actual length of the bar immediately before the occurrence of that increment. It is instructive to recognize that while stress has units of unit stress (pounds per square inch or newtons per square millimeter), unit strain has no units. It is important to repeat that unit stress has physical units, while unit strain has no units. Unit strain is strictly an abstraction unless it refers to its occurrence over a given length.

As the bar is stressed from a condition of no stress, strain increases linearly with stress until it reaches the yield stress. The slope of the initial linear portion of the curve is currently assumed to be E_s = 29,000,000 psi. To provide a perspective of its accuracy, it is relevant to know that for many years it used to be listed as 30,000,000 psi in all countries using the Imperial System of Units. When the reader converts a unit strain to unit stress using E_s and obtains 29,945.87 psi, the reader may think again about whether it is justifiable to list the answer in seven digits.

The yield stress, the stress at which the stress remains constant as strain increases, depends on the type of bar and may range from as low as 30,000 psi to as high as 180,000 psi, but in current practice the commonly used bar is the American Society for Testing and Materials (ASTM) Grade 60 bar that has a nominal yield stress of 60,000 psi. Typically, the actual yield stress is likely to be higher than the nominal value. It is very likely that the uses of ASTM Grade 80 bar (having a nominal yield stress of 80 ksi) will increase as experience with it builds up.

Ideally, stress remains constant as strain increases to a value that may range from two to ten times that at yield (Figure 4.2). Actually, there may be an upper yield stress that lasts very briefly in terms of strain. For bars Grade 60 or lower, stress starts increasing at approximately a strain of 1% at a decreasing rate with strain until the ultimate stress for the bar is reached at a strain that may range from 0.04 to 0.2

FIGURE 4.1 Reinforcing bar. (From PaulL/Shutterstock.com.)

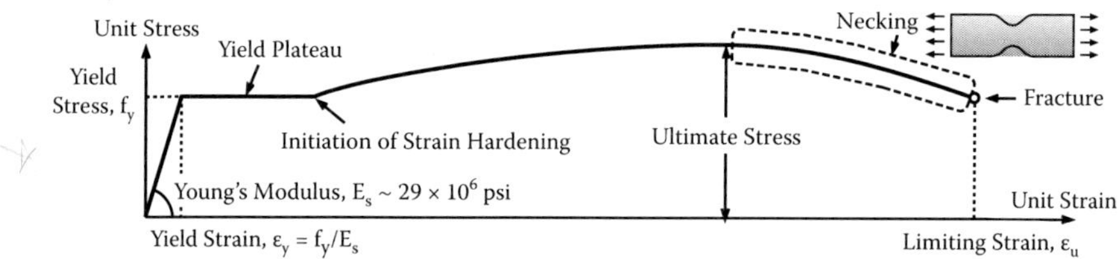

FIGURE 4.2 Idealized stress–strain characteristics of reinforcing bars.

and a stress that is approximately 50% higher than that at yield, depending on the quality and condition of the bar. At tensile strains beyond that corresponding to ultimate stress, the diameter of the bar starts to decrease perceptibly. From that point on, engineering definitions of stress and strain (stress based on original cross-sectional area and strain based on an arbitrary original length) become irrelevant to the state of the "necked" bar because the cross-sectional area of the bar is reduced and tensile deformations are concentrated in the necked region. In terms of engineering strain, fracture occurs at a value 30 to 50% higher than the strain corresponding to ultimate stress.

The stress–strain relationship in compression is approximately but not exactly the same as that in tension. In design, it is assumed to be the same.

A simple analogy provides us with a way to visualize the phenomenon represented by the stress–strain curve described. Steel is a collection of atoms. Atoms move away from or closer to one another in response to applied stress (Figures 4.3a and 4.4a), depending on the sense of the stress. If the stress does not exceed the yield stress, the atoms tend to go back to their original locations as the applied stress is reduced.

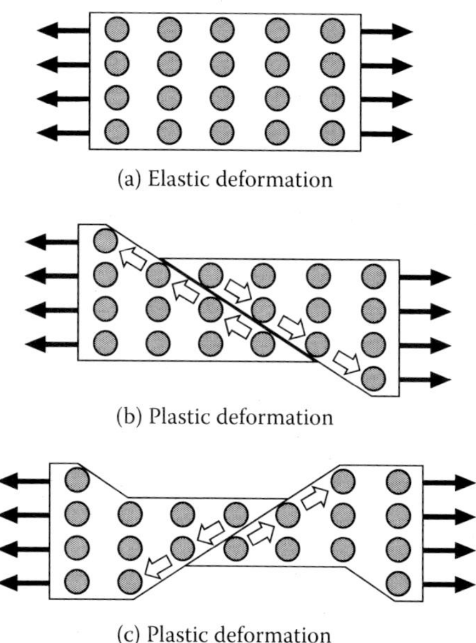

FIGURE 4.3 Atoms of steel under tensile stress.

After yielding, atoms slide on inclined surfaces as shown in Figures 4.3b and 4.4b. These relative movements of atoms result in permanent changes in the structure and dimensions of the element. If one thinks of the sliding of atoms as being similar to the sliding of an object on a surface with friction, one could conclude that the force required to continue the relative motion of atoms during yielding would remain constant after sliding starts. Because friction is usually independent of the direction of motion, one should also expect the yield stress to be similar in tension and compression. The friction analogy helps us understand the presence of permanent deformations after unloading. But the analogy fails to provide us with an explanation for strain hardening. The interaction between atoms is not captured completely by a friction model unless one can imagine an increase in friction with slip after a certain slip magnitude.

Steel used as reinforcement for concrete is commonly available as hot-rolled bars with standard geometrical and mechanical properties. The most commonly used bars are billet steel bars. These bars have fracture strains guaranteed to exceed 6% over an 8 in. gage length, including the fracture. They are produced in four "grades": 40, 60, 75, and 80. The grade refers to the specified lower bound to the yield stress. Notice that what is specified is a *lower bound*, not a mean or a median. So when we buy Grade 60 billet steel, the most widely used reinforcing steel, it is very unlikely that the actual yield stress is 60 ksi. In most cases, the actual yield stress exceeds the nominal value, the best estimate of the mean yield stress ranging usually from 65 to 75 ksi.

In design, we assume that the yield stress, f_y, of Grade 60 bars is 60 ksi.

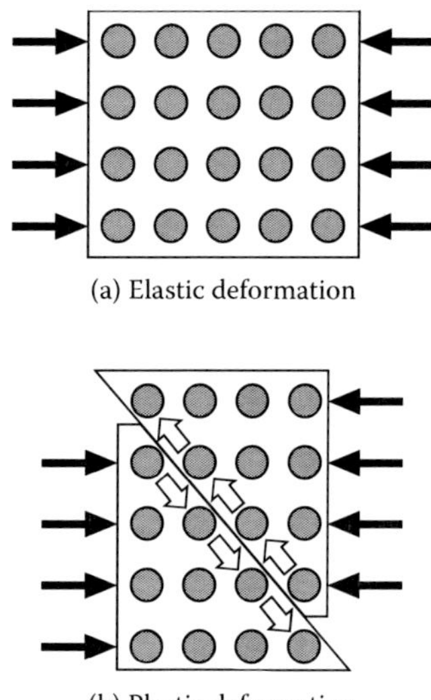

(a) Elastic deformation

(b) Plastic deformation

FIGURE 4.4 Atoms of steel under compressive stress.

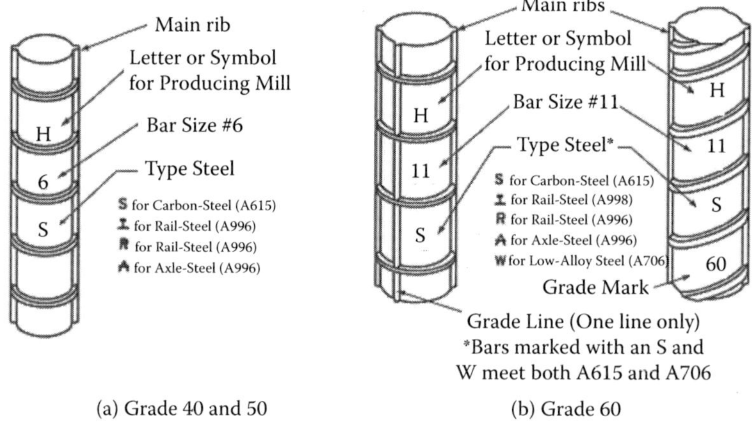

(a) Grade 40 and 50

(b) Grade 60

FIGURE 4.5 Bar marks. (From the Concrete Reinforcing Steel Institute, http://www.crsi.org/index.cfm/steel/identification.)

Billet steel bars should not be welded because welding makes them brittle. If welding is required, low-alloy steel bars should be used. Low-alloy steel bars typically have yield points between 60 and 68 ksi and elongations at rupture exceeding 10% over an 8 in. gage length.

In the United States practice bars are identified by a pound sign (#) and the size of the nominal diameter in eighths of an inch (1/8 in.). For example, a #8 bar is a bar with a nominal diameter of 1 in. Sizes and nominal yield stresses of bars #3 to #18 are marked as shown in Figure 4.5.

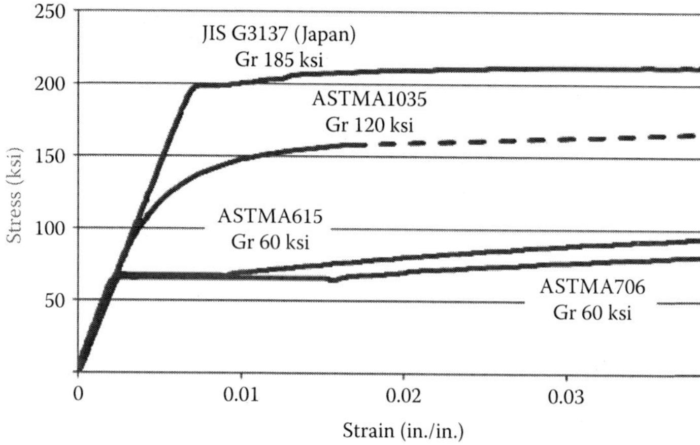

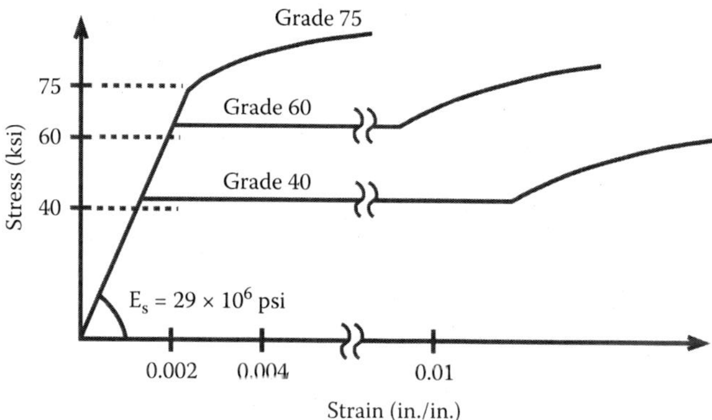

FIGURE 4.6 Unit stress vs. unit strain for bars of different grade.

The unit stress–unit strain curves of reinforcing steel bars of different grades have different shapes (Figure 4.6). The elongation at rupture is also different between one grade and another. In general, steels with higher strengths tend to have shorter yield plateaus (if any) and smaller deformations at rupture.

Table 4.1 lists standard bar sizes, cross-sectional areas, and weights per foot of length. The bars we use today have surface deformations to improve their bond with concrete. They are not prismatic, and their actual cross-sectional areas listed deviate from the areas of circles with diameters equal to the listed diameters. For these reasons, we refer to the dimensions in Table 4.1 as nominal dimensions. It is of interest to note that the unit weights in the table are consistent with listed diameters. For deformed bars, the listed cross-sectional areas are essentially nominal.

Standard bars are sold in 20, 40, and 60 ft lengths. Handling at the construction site of individual bars weighing more than approximately 90 lbf is considered to be difficult.

TABLE 4.1
Properties of Standard Bars

Bar Size	Diameter (in.)	Area (in.^{2})	Nominal Unit Weight (lbf/ft)
3	0.375	0.11	0.376
4	0.50	0.20	0.668
5	0.625	0.31	1.043
6	0.75	0.44	1.502
7	0.875	0.60	2.044
8	1.00	0.79	2.67
9	1.128	1.00	3.40
10	1.27	1.27	4.303
11	1.41	1.56	5.313
14	1.693	2.25	7.65
18	2.257	4.00	13.6

EXAMPLE 4.1

Consider a 10 ft long Grade 60 billet steel #8 bar subjected to tensile axial force. Draw an approximate relationship between force and elongation.

SOLUTION

The yield stress of the bar is expected to be between 65 and 75 ksi. The cross-sectional area is 0.79 in.^{2}. Therefore, the force that makes the bar yield is expected to be between $65 \times 0.79 = 51$ kip and $75 \times 0.79 = 59$ kip. This is approximately equal to the weight of ten standard pickup trucks.

With Young's modulus assumed to be 29×10^3 ksi, the yield strain should be ideally between $65/(29 \times 10^3) = 0.0022$ and $75/(29 \times 10^3) = 0.0026$.*

The length of the bar is 10 ft = 120 in. Therefore, the elongation at which the bar yields is between 0.0022×120 in. = 0.26 in. and 0.0026×120 in. = 0.31 in.

Strain hardening (on average) is likely to occur at an elongation of approximately $0.01 \times 120 = 1.2$ in. The force-elongation relationship should be within the shaded region shown in Figure 4.7 after yielding.

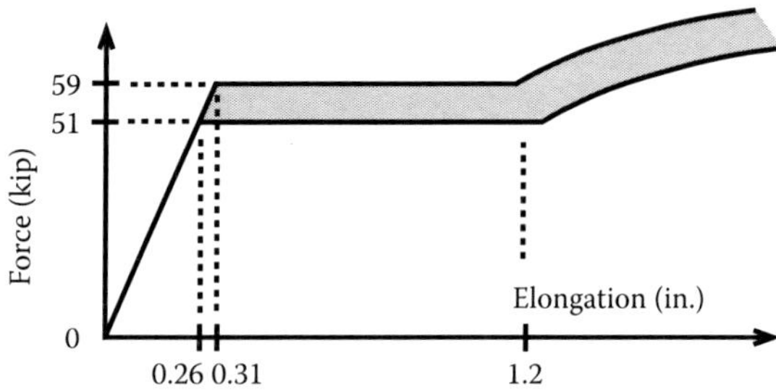

FIGURE 4.7 Force–elongation relationship of a 10 ft long Grade 60, A706, #8 bar.

* Some material specifications allow the strain at yield for a 60 ksi bar to be as much as 0.005.

BARE ESSENTIALS

- Currently bar quality is identified by the "grade" of the bar, e.g., "Grade 60" refers to a bar with a minimum yield stress of 60 ksi.
- Bar size is identified by a number that is approximately equal to the nominal diameter in eighths of an inch.
- Although in conceptual design the stress–strain relationship of reinforcing bars is assumed to be "elasto-plastic," actually the bar may develop a stress ~1.5 times the yield stress if strained beyond the yield strain.
- It would not be surprising for a Grade 60 bar to have a maximum stress of 90 ksi at a unit engineering strain of ~0.1. After the strain at maximum stress is exceeded the bar diameter starts decreasing and the definition of engineering strain provides only an approximate value of the strain.
- Fracture strain for a Grade 60 bar is specified to be as low as 0.06.
- Bars with yield stresses higher than those of Grade 60 bars tend to have smaller strains at fracture.

EXERCISES

1. Repeat the example for a 20 ft long Grade 60 billet steel #11 bar subjected to tensile axial force.
2. Two 40 ft long bars are going to be lifted using a crane (one bar at a time). Both bars are Grade 60 bars. One is a #6 bar and the other is a #11 bar. The crane has a spreader beam that allows the crane operator to lift the bars from two points. Assuming that forces applied to the bar by the lifting rig are vertical, determine the locations of the points where the lifting rig should be attached to the bars in order to avoid yielding of the bars during lifting.
3. Ignoring strain hardening (a typical assumption in design) and assuming $f_y = 60$ ksi, compute the stresses associated with the following strains:
   Tensile strains: 0.001, 0.002, 0.0021, 0.003, 0.01, 0.02.
   Compressive strains: 0.001, 0.002, 0.0021, 0.01.

5 Concrete

In this section, we focus on the short-time mechanical properties of concrete that relate to its function as a building material. By "short time" we refer to a loading period that is typically more than a minute but not exceeding a day. Concrete is a conglomerate obtained by mixing Portland cement, a coarse aggregate (typically gravel or crushed rock), a fine aggregate (sand), and water. Cement is the ingredient that, with the help of water, hardens and transforms the mixture into a synthetic rock. In addition to cement, chemical and natural admixtures can be used to improve the workability of the wet mix and to increase the strength, durability, and fire resistance of the hardened concrete. Admixtures may also be used to control the tendency of the concrete to change volume with time.

Figure 5.1 shows a saw cut through hardened concrete. The dark objects seen are pieces of coarse aggregate. The grey matrix is the mortar, a mix of cement paste and sand. Concrete is usually made using natural rock (normal weight aggregate) and weighs 140 to 150 pcf. Its density may be reduced by using lightweight aggregate such as pumice. A subtle but important property of concrete is its rate of thermal expansion, which is very close to that of steel (approximately 0.000006 per deg. Fahrenheit for temperatures not exceeding approximately 700 deg. Fahrenheit).

5.1 COMPRESSIVE STRENGTH

Mechanical properties of concrete depend on the proportions of the mix and the characteristics of coarse and fine aggregates. The defining property of concrete is its compressive strength at 28 days, f'_c. In the design environment, the stiffness of the concrete and its tensile, shear, and bond strengths are expressed traditionally, in the United States, in terms of its compressive strength determined from tests of concrete cylinders with a diameter of 6 in. and a height of 12 in. loaded to failure in approximately two minutes. Figure 5.2 contains a series of photographs taken during the test of a cylinder. In that test, failure of the cylinder in compression was preceded by longitudinal cracks parallel to the axis of the cylinder. Such cracks are attributed to the lateral strains developed in the concrete as it is compressed axially.

Representative short-time stress–strain curves for three different concrete strengths are shown in Figure 5.3. It is observed that, up to a stress of half the strength, the increase in stress is approximately proportional to strain. As the applied stress is increased, nonlinear response is initiated and the material softens progressively until the maximum stress is reached at a strain ranging from 0.0015 to 0.0025, depending on strength. Typically, longitudinal cracks are visible at a load ranging from 0.7 to 0.8 f'_c as illustrated ideally in Figure 5.3c. Disintegration of the cylinder may occur

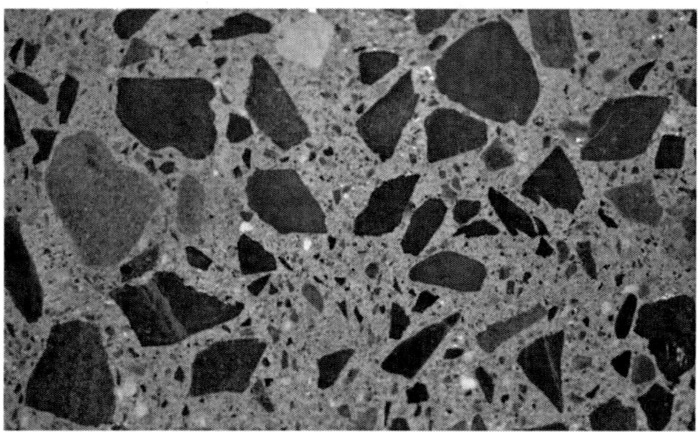

FIGURE 5.1 Saw cut through hardened concrete.

FIGURE 5.2 Test of a cylinder.

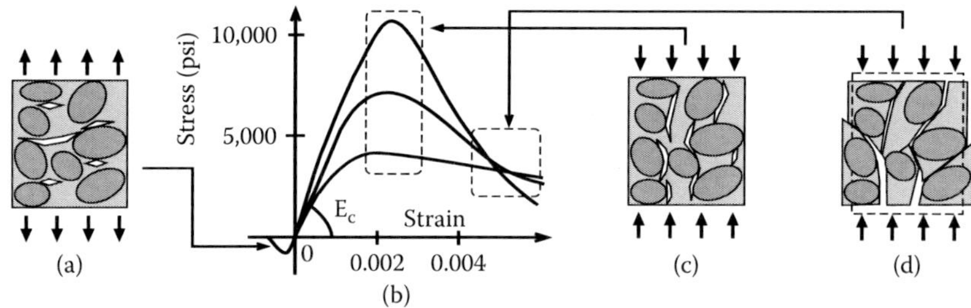

FIGURE 5.3 (a) Propagation of tensile cracks. (b) Stress–strain relationship. (c) Vertical cracking by compressive stress. (d) Sliding failure.

through widening of some of these longitudinal cracks or sliding along inclined cracks (Figure 5.3d). The strain at failure is difficult to establish, primarily because the integrity of the cylinder beyond maximum stress depends on the stiffness of the loading system. It is evident from the shape of the stress–strain relationship that if the cylinder is loaded by weights applied at its top surface, equilibrium would not be satisfied immediately after strength reduction starts. In practice, the limiting compressive strain of concrete is assumed to be 0.003. In considerations related to investigating behavior of reinforced concrete elements, the limit is commonly set at 0.004. Under special conditions of loading it may be set at higher levels.

Various professional engineering communities use different statistical procedures to obtain a nominal value for design strength from a set of cylinder test results. The

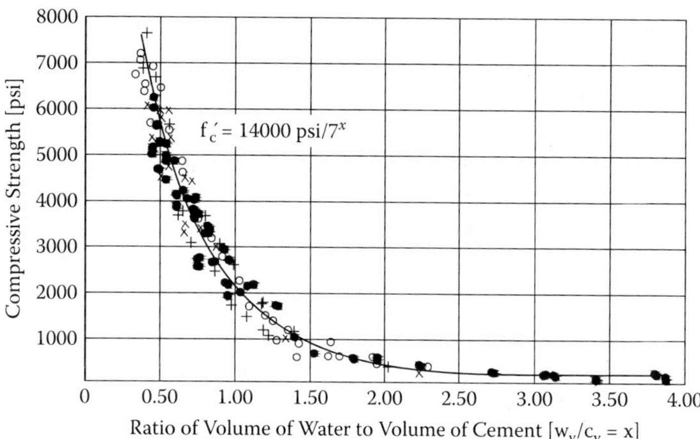

FIGURE 5.4 Effect of water-to-cement ratio on compressive strength.

nominal design strength tends to be set at approximately 1000 psi below the mean strength obtained from the cylinder tests.

With respect to strength, the ratio of water to cement in the mix is the most important parameter. The lower this ratio, the higher is the compressive strength (Figure 5.4). Though it may seem obvious in retrospect, it took a long time and many tests by Duff Abrams (1918) to discover the dependency of compressive strength primarily on the water-to-cement ratio. His discovery was unquestionably the most important contribution to concrete technology. Above all, Abrams, a student of Talbot, had the right engineering perspective. He ignored the secondary effects and summarized his discovery in a very simple expression:

$$f'_c = 14{,}000/7^{w_v/c_v} \tag{5.1}$$

where f'_c is the compressive strength at 28 days in psi for normal weight aggregate concrete, w_v is the volume of water in mix, and c_v is the volume of cement in mix.

5.2 STIFFNESS

Stiffness or Young's modulus, E_c, for concrete is assumed to be the initial slope of the stress–strain curve obtained in the cylinder test (Figure 5.3). The first approximation for Young's modulus in terms of compressive strength for normal weight aggregate concrete was simply

$$E_c = 1000\, f'_c \tag{5.2}$$

where E_c is Young's modulus (psi) and f'_c is the compressive strength (psi).

Equation 5.2 continues to be a good tool for making initial estimates in design. The initial stiffness of the concrete has been recognized to be sensitive to changes in compressive strength. Actually, it is sensitive to many variables, such as relative volumes

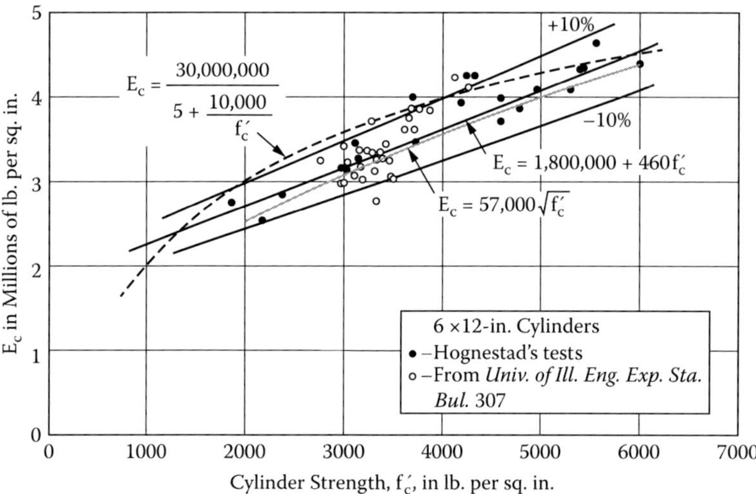

FIGURE 5.5 Relationship between Young's modulus and cylinder strength.

of paste and aggregate, stiffnesses of the aggregates, and loading rate. Design expressions for Young's modulus of concrete have changed over the years, but most have considered explicitly only compressive strength and, in one case, density. Examples are given in Equations 5.3 to 5.5 and Figure 5.5, along with specific measurements reported by Hognestad. Equations 5.3 to 5.5 are listed in chronological order.

$$E_c = 1,800,000 + 460\, f'_c \tag{5.3}$$

$$E_c = 30,000,000/(5 + 10,000/f'_c) \tag{5.4}$$

$$E_c = 33 w_c^{1.5}\sqrt{f'_c} \tag{5.5}$$

where E_c is Young's modulus (psi), f'_c is the compressive strength of concrete in psi (from 6 × 12 in. cylinders), and w_c is the unit weight of concrete (pcf).

For normal weight aggregate concrete, assumed to have a density of 145 pcf, Equation 5.5 leads to

$$E = 57,000\sqrt{f'_c}\text{, with } E_c \text{ and } f'_c \text{ in psi} \tag{5.6}$$

5.3 TENSILE STRENGTH

Although it is intimately involved in all aspects of the strength of reinforced concrete, tensile strength is seldom included explicitly in engineering calculations. Tensile strength of concrete is difficult, if not impossible, to measure directly because it is difficult to load a tension sample concentrically and independently of local stresses at the grips. Direct tensile strength of normal weight aggregate concrete is generally estimated to be approximately

$$f_t = 4\sqrt{f'_c} \tag{5.7}$$

where f_t is the direct (axial) tensile strength of concrete (psi) and f'_c is the compressive strength of concrete (psi).

The relationship between compressive and tensile strengths is also sensitive to the surface characteristics of the coarse aggregate, a variable that is missing in Equation 5.7 for the understandable reason that the scatter in the data justifies the simplest approach.

Tensile strength is measured indirectly by two popular test types. One is the splitting or the Brazilian test shown in Figure 5.6a. The other is the modulus of rupture test.

In the splitting test, a concrete cylinder is, in effect, squeezed by two equal and opposite line loads applied along the height and perpendicular to a diameter of the cylinder. The test is terminated by a crack along the loaded diameter (Figure 5.6a). An element in the cylinder is stressed biaxially, as shown in Figure 5.6b. The tensile strength is determined using an expression derived assuming linear response:

$$f_{ts} = 2P_s/(\pi D h_c) \tag{5.8}$$

where f_{ts} is the tensile strength (psi), P_s is the maximum force reached in test, D is the diameter of the test cylinder, and h is the height of the test cylinder.

Equation 5.8 refers to a homogeneous material in the range of linear response. The value obtained from it using the load and dimensions from the test is assumed to represent a good estimate of the direct tensile strength of concrete and is typically lower than the modulus of rupture.

The modulus of rupture involves the test of a simply supported beam (usually measuring 6 × 6 in. in cross section) subjected to two concentrated loads symmetrically located about the center of the beam span. The modulus of rupture is calculated from the measured maximum moment assuming that the section of the beam responds linearly at failure.

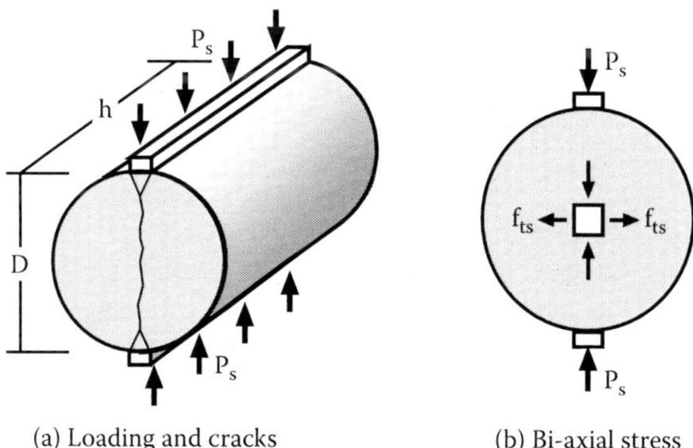

FIGURE 5.6 Splitting test.

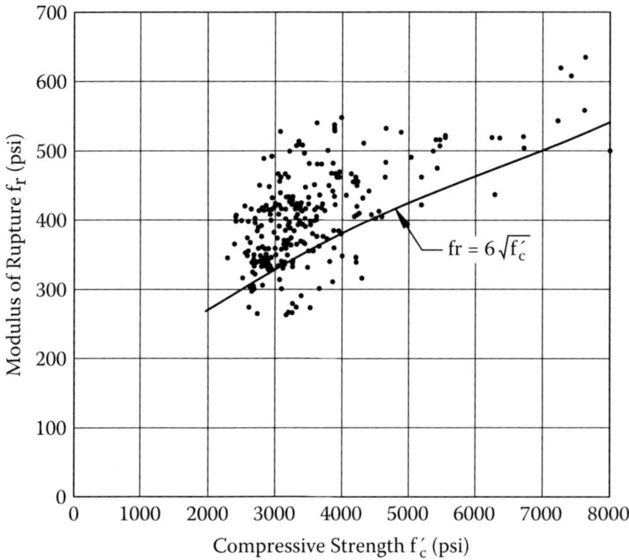

FIGURE 5.7 Relationship between modulus of rupture and compressive strength.

$$f_r = 6M_r/b^3 \tag{5.9}$$

where f_r is the modulus of rupture (Figure 5.7), M_r is the maximum moment reached in test, and b is the width of the square section.

For normal weight aggregate concrete, the expression

$$f_r = 6\sqrt{f'_c} \tag{5.10}$$

where f_r is the modulus of rupture (psi) and f'_c is the compressive strength (psi), provides a reasonable lower bound to the modulus of rupture (Figure 5.7). The modulus of rupture is approximately 50% higher than the corresponding tensile strength because the response of concrete at fracture is not linear, as assumed in the derivation of Equation 5.9.

5.4 A FORMULATION FOR THE STRESS–STRAIN RELATIONSHIP OF CONCRETE

Using data from tests of eccentrically loaded columns made with concretes with strengths not exceeding 6000 psi and following ideas published by Ritter, Hognestad proposed the following expression for the stress–strain relationship for concrete in compression:

$$f_c = f''_c\left[2\frac{\varepsilon_c}{\varepsilon_o} - \left(\frac{\varepsilon_c}{\varepsilon_o}\right)^2\right] \quad \text{if } \varepsilon_c \le \varepsilon_o \tag{5.11}$$

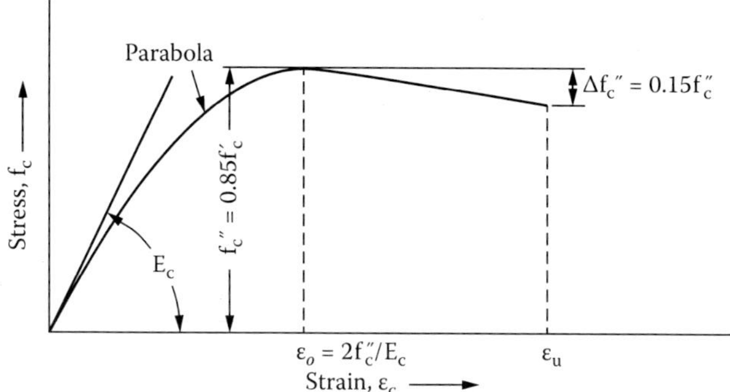

FIGURE 5.8 Stress–strain relationship proposed by Hognestad.

$$f_c = f_c''\left[1 - \frac{\varepsilon_c - \varepsilon_o}{0.004 - \varepsilon_o} \cdot 0.15\right] \text{ if } \varepsilon_c \geq \varepsilon_o \tag{5.12}$$

where E_c is the modulus of elasticity (psi), f_c is the concrete stress (psi), f_c'' is the peak stress, ε_c is the concrete strain, and ε_o is the concrete strain at peak stress (approximated as $2f_c''/E_c$).

This expression is plotted in Figure 5.8. It will be used in the following sections in determining strength of reinforced concrete sections subjected to bending and axial load. Hognestad limited the maximum stress (f_c'') to 85% of the cylinder strength for reasons to be discussed in Chapter 7.

EXAMPLE 5.1

Assume we are asked to design a 5-story office building with nine columns having square cross sections (Figure 5.9) to be built with concrete of nominal strength = 5000 psi. We choose the limit to mean compressive stress in all the columns for service level loads equal to 20% of the specified cylinder strength (Carneiro, 1947). We assume that the dead and live loads per unit area are 200 and 30 psf, respectively. We set out to determine (1) the required cross-sectional dimension and (2) the total shortening of the interior column in the first story.

Before we start the exercise, we take a moment to think. What is the result of our calculation likely to be? In classical architecture, columns were proportioned to mimic the proportions of the human body. That would suggest that the width of the square column should be approximately one-seventh of the clear height, or 20 in. (Chapter 2). When we finish our calculations, it would be wise to see how the dimension obtained by calculation compares with our initial guess.

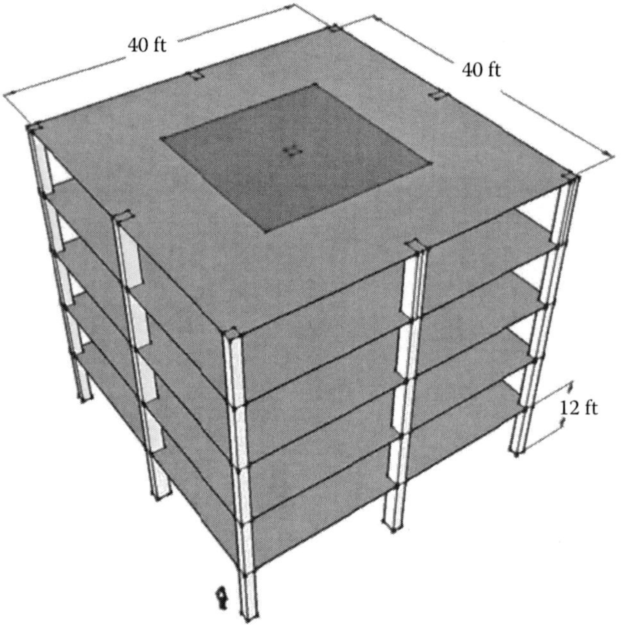

FIGURE 5.9 Five-story office building with nine columns.

SOLUTION

1. The weight of the building (including dead and live loads) is

$$W = 5 \times 40 \text{ ft} \times 40 \text{ ft} \times (200 \text{ psf} + 30 \text{ psf}) = 1{,}840{,}000 \text{ lbf}$$

We assume that the axial force carried by the interior column in the first story is proportional to a tributary area defined to be 20 × 20 ft, as shown in dark grey in Figure 5.9 (the assumed tributary area corresponds to one-fourth of the total floor area):

$$W = 1{,}840{,}000 \text{ lbf}/4 = 460{,}000 \text{ lbf}$$

To make the average compressive stress less than 5000 × 1/5 = 1000 psi, the cross-sectional area of each column shall be more than

$$A_g = 460{,}000 \text{ lbf}/1000 \text{ psi} = 460 \text{ in.}^2$$

Therefore, the depth of the section shall be more than

$$h = \sqrt{(460 \text{ in.}^2)} \approx 21 \text{ in.}$$

Because of constructors' preferences, the recommended depth should be expressed as a multiple of 2 in. We select a depth of 22 in.

2. The mean compressive stress is approximately

$$f = P/A_g = 460{,}000\ \text{lbf}/(22 \times 22\ \text{in.}) = 950\ \text{psi}$$

We have ignored the contribution of the reinforcement because including it would not improve the quality of the result given the approximations involved in the other steps of our solution.

Expressed with no more than two significant figures, the modulus of elasticity is approximately

$$E_c = 57{,}000\sqrt{5000}\cdot \text{psi} = 4\cdot 10^6\ \text{psi}$$

Because the calculated stress is much smaller than the compressive strength, we assume that the strain is

$$\varepsilon = f/E_c = 950\ \text{psi}/(4.0 \times 10^6) = 0.00024$$

At this strain, the stress in the reinforcement would be

$$0.00024 \times 29 \times 10^3\ \text{ksi} = 7\ \text{ksi}$$

If 1% of the cross-sectional area is occupied by steel reinforcement, the contribution of the reinforcement to the resistance of the column is

$$7\ \text{ksi} \times 0.01 \times (22\ \text{in.})^2 = 34{,}000\ \text{lbf}$$

or 7% of the estimated load. Again, this is a tolerable error given the nature of our calculations.

Multiplying the strain and the length of the column (12 ft × 12 in./ft = 144 in.) yields an estimate of the short-time shortening of the column under its total design load:

$$\delta = \varepsilon \times H = 0.00024 \times 144\ \text{in.} \approx 1/32\ \text{in.}$$

A column with 5000 psi concrete may be assigned a working stress of 1800 psi. In that case, the calculated short-time deformation would be 1/16 in. If the axial load is permanent, the deformation may triple with time (Chapter 6).

BARE ESSENTIALS

1. Tensile strength of concrete is assumed to be proportional to the square root of compressive strength and is approximately 1/10 of compressive strength.
2. Young's modulus of concrete is assumed to be proportional to the square root of compressive strength and varies between 1/10 and 1/5 that of steel.
3. The unit weight of concrete made with normal weight aggregate is approximately 150 pcf.
4. For normal weight aggregate concrete, compressive strength is expected to be approximately 5000 psi at an age of 28 days for a water-to-cement ratio of 0.5.

EXERCISES

1. Repeat the example above assuming that the building has 10 stories.
2. Compute the ratio of the product $0.75 \times f'_c \times 0.003$ to the area under the stress–strain curve defined by Equations 5.11 and 5.12 between $\varepsilon_c = 0$ and $\varepsilon_c = 0.003$. Assume $f'_c = f''_c$ and
   a. $f'_c = 3000$ psi
   b. $f'_c = 4000$ psi
   c. $f'_c = 5000$ psi

   What can you conclude from your results? Note: Recall that the definition of average is

$$\left(\int_a^b f(x)dx\right) \div (b - a)$$

3. Consider a normal weight concrete beam with a span-to-depth ratio of 12 and simple supports. Estimate the maximum length that the beam could span without cracking under its own weight. Ignore variations in the tensile strength of concrete with time. Assume $f'_c = 5000$ psi.

6 Time-Dependent Volume Changes of Concrete

Shrinkage and Creep

Concrete changes its volume with time. The magnitude and rate of the change depend on the relative humidity of the environment and the applied stress as well as the properties of the concrete. Time-dependent volume changes are easier to understand if we consider them in only one direction and in an environment of constant stress and relative humidity.

Consider a concrete cylinder immediately after setting. The axial stress caused by self weight may be ignored. In a dry environment, the height of the cylinder will shorten with time (as will the diameter). This shortening is attributed primarily to loss of adsorbed water. The loss sets up internal stresses that lead to compressive strain or shortening of the cylinder. The change in strain is higher near the surfaces of the cylinder. If we ignore that and plot the ratio of the shortening of the cylinder against time, the relationship observed is described by Figure 6.1. The shortening is rapid initially but slows down with time. In an environment of constant moderate humidity, more than half of the total shortening occurs within three months. Shortening of the cylinder without externally applied force is called *shrinkage*.

Consider the same cylinder with an applied axial force at a given age t_0. Assuming that the applied force generates a compressive stress, f_c, below half the strength of the concrete, the measured change in strain would be approximately $\varepsilon_{inst} = f_c/E_c$, where E_c is Young's modulus. If the force is sustained and strain measurements are continued beyond time t_0, the curve to be obtained, with the shrinkage strain subtracted from the results, will look like the one in Figure 6.2. The portion of the time-dependent strain related to applied stress is called *creep strain*. It varies with time in a manner similar to the variation of shrinkage, but it does not start immediately after concrete is set. Creep starts after the cylinder is subjected to an external stress. Total magnitude of creep strain is usually expressed as a multiple m of the instantaneous strain, ε_{inst}, caused by the applied stress.

Both shrinkage and creep depend on many factors. What we know about them comes from experience and experiments. In the following, we shall refer to the approach adopted by the European Concrete Committee to note the main factors that affect the rate and magnitude of time-dependent volume changes (CEB-FIP, 1973). As we discuss the numbers offered by the European Committee, the reader should keep in mind that the best source of information for a structural engineer is the observed behavior of concrete in buildings built using the same materials and under similar environmental conditions as the one he or she intends to build. If the student is ever

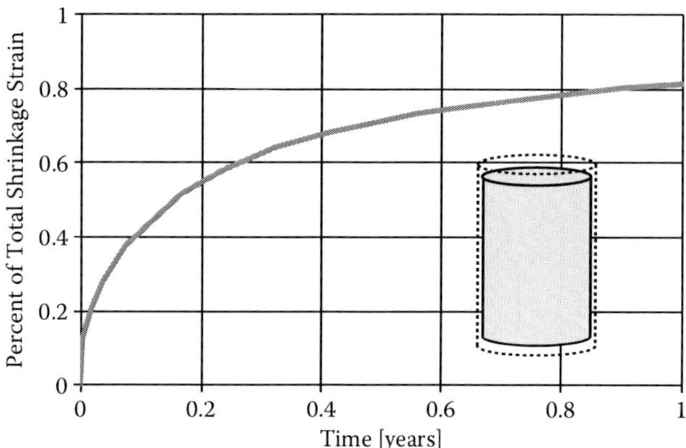

FIGURE 6.1 Variation of shrinkage strain over time.

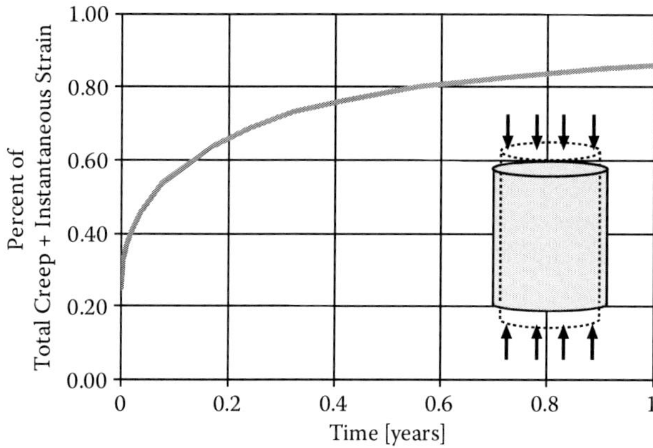

FIGURE 6.2 Variation of creep strain over time.

required to estimate volumetric changes of concrete, he or she should remember that, in general, a reasonable upper bound to shrinkage unit strain is 0.0005, and the additional strain caused by creep is likely to be from one-half to three times the instantaneous strain. If the environmental humidity remains at approximately the same level, 80% of the total expected change occurs within one year after setting for shrinkage and one year after loading for creep.

6.1 SHRINKAGE

The main variables affecting shrinkage are as follows:

- Humidity
- Volume-to-surface ratio (the shape of the cross section of the structural element)
- Amount of cement
- Water-to-cement ratio

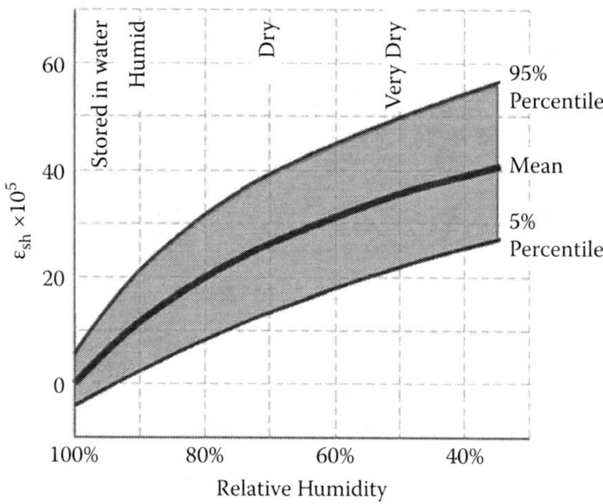

FIGURE 6.3 Variation of shrinkage strain with relative humidity.

Figure 6.3 shows the variation of total shrinkage strain (ε_{sh}) with humidity. The data plotted were obtained for

- Plain concrete made with 600 to 750 lb of cement per yd^3
- Concrete at room temperature
- A water-to-cement ratio of approximately 0.45 (shrinkage increases with increasing water-to-cement ratio; shrinkage strains for a water-to-cement ratio of 0.6, being close to 50% higher than those given)

The reader should realize that the ranges within which the data shown are applicable are narrow. And yet, within those ranges, the spread of the data is large. In general, a reasonable upper bound to shrinkage unit strain is 0.0005. Shrinkage can be controlled to be low by using the proper admixture and curing process.

6.2 CREEP

Creep is quantified as the ratio of *additional* strain (caused by applied stress) to instantaneous strain ε_{inst} at stress levels below half the compressive strength. This ratio is called creep coefficient m, and it too is sensitive to relative humidity (Figure 6.4).

We note that the value of the creep coefficient varies from approximately 0.5 to approximately 3. At 70% relative humidity (within the 5th to 95th percentile limits) it ranges from 1¼ to 2.

Creep varies with thickness and water-to-cement ratio, as in the case for shrinkage. The values shown in Figure 6.4 apply to

- Plain concrete made with 600 to 750 lb of cement per cubic yard and loaded two to six weeks after it is cast
- Concrete at room temperature

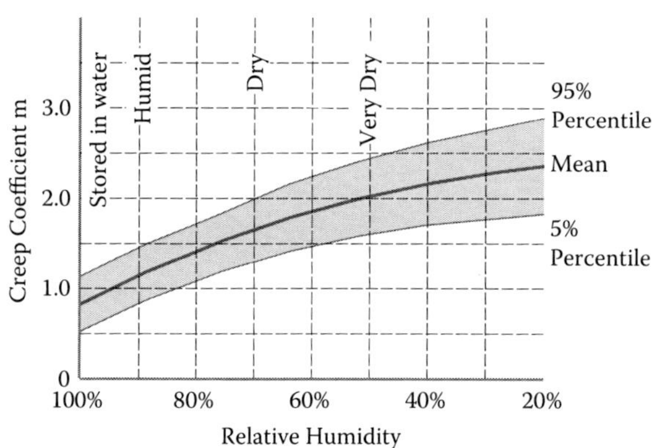

FIGURE 6.4 Variation of creep coefficient with relative humidity.

- A ratio of the volume to the surface that is approximately equal to 6 in. (the creep coefficient increases with decreasing ratio of volume to surface, with creep coefficients at a volume-to-surface ratio of 3 in. being approximately 6/5 of those given).
- A water-to-cement ratio of approximately 0.45 (creep increases with increasing water-to-cement ratio, with creep strains for a water-to-cement ratio of 0.6 being close to 60% higher than those given).

The important conclusion we make about the creep strain is that it too is affected by the relative humidity, the shape and size of the concrete element, the cement content, and the water-to-cement ratio. In addition, its magnitude is sensitive to age at loading. Usually, the creep coefficient does not exceed 3.

It is also important to know that, although it is proportional to the initial applied stress, the additional strains caused by creep do not disappear when the stress is removed. This is illustrated in Figure 6.5. The total deformation in an element loaded over a long period of time is the product of the initial deformation and (1 + m). When the element is unloaded, part of the total deformation is recovered immediately, another part is recovered over time, and the rest may be permanent. For additional information on creep, refer to the work by A.D. Ross (1958).

6.3 SHRINKAGE AND CREEP VS. TIME

The rate of change with time of shrinkage and creep is represented by a coefficient ρ_t, which indicates the strain at a given time as a fraction of the total strain.

It is interesting to observe that, according to the experience summarized in Figure 6.6, one cannot make the claim that the time-dependent volume change ever stops. We also note that almost 90% of the time-dependent strain should occur within the first two years (if there is no strong change in humidity and, in the case of creep, in load).

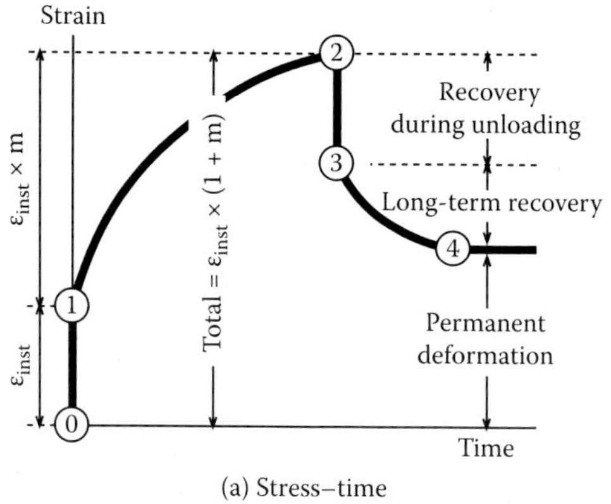

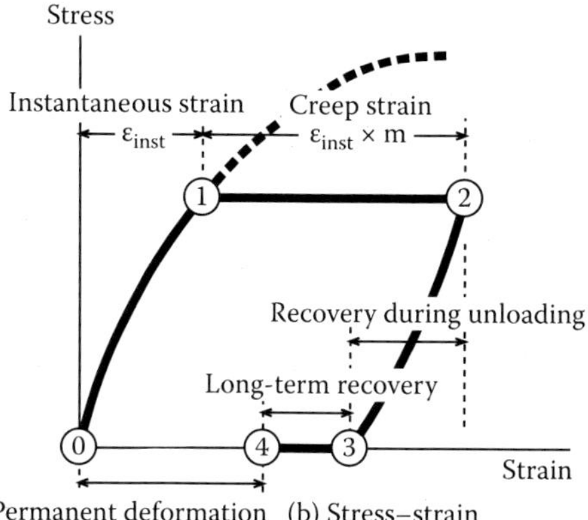

FIGURE 6.5 Variation of deformations in concrete over time.

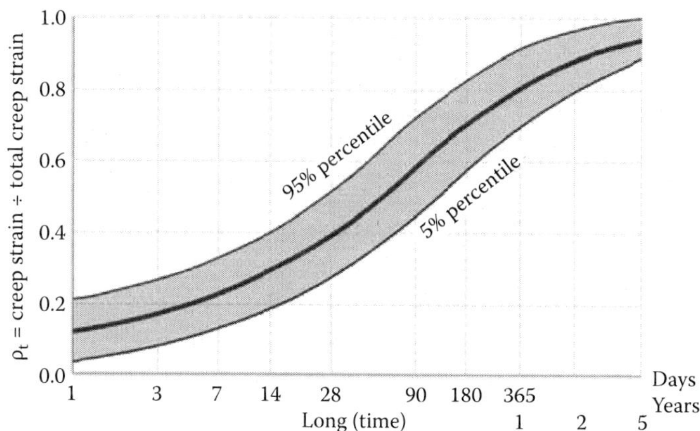

FIGURE 6.6 Variation in shrinkage and creep deformations over time.

EXAMPLE 6.1

A 12 ft long column with a 22 × 22 in. cross section supports a load of 200 kips (Figure 6.7). Estimate the order of magnitude of the total shortening of the column after five years. Assume E_c = 4000 ksi.

SOLUTION

Ignoring the effect of steel reinforcement, the applied stress is

$$P/A = 200\ \text{kip}/(22 \times 22\ \text{in.}) = \sim 410\ \text{psi}$$

The instantaneous strain is

$$\varepsilon_{inst} = \frac{f_c}{E_c} = \frac{P/A}{E_c} = \frac{200\ \text{kip}/(22\ \text{in.} \times 22\ \text{in.})}{4000\ \text{ksi}} \approx 0.0001^{*}$$

We assume that the shrinkage strain is not likely to exceed 0.0005 by much, so we assume

$$\varepsilon_{sh} = 0.0005$$

We assume the creep coefficient to be

$$m = 3$$

After five years most of the long-term deformations are likely to have taken place ($\rho_t = 1$) (Figure 6.6):

$$\varepsilon_{total} = \varepsilon_{inst} + \rho_t \times (\varepsilon_{sh} + m \times \varepsilon_{inst}) = 0.0001 + 1 \times (0.0005 + 3 \times 0.0001) = 0.0009.$$

The total shortening is therefore likely to be on the order of $\delta = 0.0009 \times 144$ in. = ~1/8 in.

---

* If the section had, for instance, a reinforcement ratio $A_{st}/A_g = 1\%$, the effect of the reinforcement would reduce this strain to

$$\varepsilon_{inst} = \frac{P}{E_c \cdot \left(A_g + (n-1) \cdot A_{st}\right)} = \frac{200\text{kip}}{4000\ \text{ksi} \times (22\ \text{in.} \times 22\ \text{in.}) \times (1 + (7.3 - 1) \times 0.01)} = 0.000097$$

A_g is the gross cross-sectional area, and A_{st} is total cross-sectional area of longitudinal steel. The parameter n is the ratio of the modulus of elasticity of steel to the modulus of elasticity of concrete. The relative error made ignoring the effect of the steel on instantaneous strain is negligible (and equal to $(n - 1) \times A_{st}/A_g$) given the other approximations made.

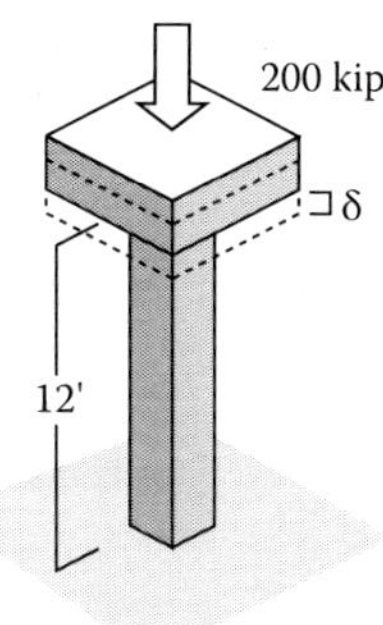

FIGURE 6.7 Example 6.1.

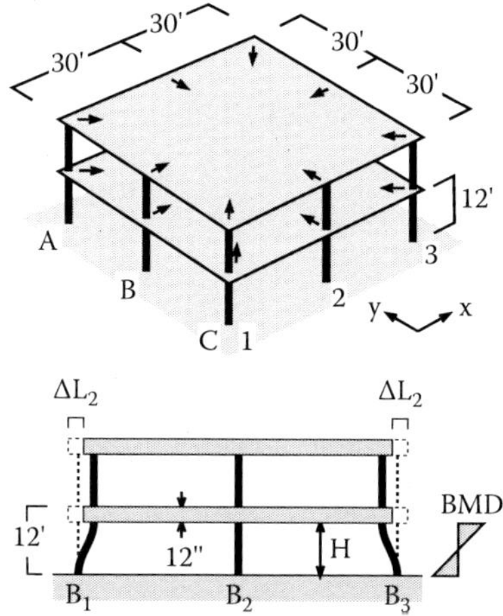

FIGURE 6.8 Example 6.2 (deformations have been exaggerated).

EXAMPLE 6.2

The overall dimensions of a 2-story parking structure are shown in Figure 6.8. The floor slabs are 12 in. thick. The compressive strength of concrete is $f'_c = 4000$ psi. The columns measure 20 × 20 in. in cross section. We are asked to estimate the forces in first-story columns B1, B2, and C1 that may be induced by shrinkage of the slabs.

SOLUTION

We understand that there is no exact answer to the question. Shortening of the slabs depends on the shrinkage of the concrete, the environmental conditions, the amount of slab reinforcement, and the stiffness of the columns. To get a reasonable upper bound we ignore the restraints imposed on slab shortening by the slab reinforcement and column stiffness and assume a reasonable upper bound to the shrinkage unit strain ε_{sh} of 0.0005.

Under ideal circumstances, column B2 is not likely to be affected by the shrinkage of the slab because it is located in the middle of the structure. The top of column B1 will be pulled toward the center. If we assume that the structure is truly symmetrical, the expected shortening of the distance between columns B1 and B2 will be approximately

$$\Delta L_2 = \varepsilon_{sh} \times L_2 = 0.0005 \times 30 \text{ ft} \times 12 \text{ in./ft} = \sim 0.2 \text{ in.}$$

where L_2 is the center-to-center distance between columns B1 and B2.

The calculated shortening would generate forces on the columns depending on their stiffnesses. To obtain an upper bound, we ignore the possibility of cracking in the columns (a phenomenon that would reduce their stiffnesses), but we reduce Young's modulus for concrete to account for creep:

$$E_c = \frac{1}{3} \cdot 57{,}000\sqrt{f_c' \cdot \text{psi}}$$

We assume the footings and the slab to restrain rotations at their supports so that the column stiffness for lateral loading is

$$K = \frac{12 \cdot E_c \cdot I_g}{H^3} = \frac{12 \cdot \frac{1}{3} \cdot 57{,}000\sqrt{f_c' \cdot \text{psi}} \cdot \frac{1}{12} \cdot (20 \text{ in.})^4}{(144 \text{ in.} - 12 \text{ in.})^3} = 83.6 \frac{\text{kip}}{\text{in.}}$$

Here H is the clear height of columns and I_g is the moment of inertia of the gross cross section (computed as $b^4/12$, with b = 20 in. being the side dimension of the square column section).

The estimated shear on column B1 is $V_{B1} = K \times \Delta L_2 = \sim 83.6$ kip/in. × 0.18 in. = ~15 kip.

We now consider column C1. Because the spans in directions x and y are the same, the column moves approximately 0.2 in. in both direction x and direction y (Figure 6.8). The components of the shear force on the column in directions x and y are both equal to 15 kip. Therefore, the estimated shear on column C1 is

$$V_{C1} = \sqrt{(15 \text{ kip})^2 + (15 \text{ kip})^2} = 21 \text{ kip}$$

It is of interest to point out that the shear force of 21 kip has relatively little meaning on its own. It is always useful to convert shear forces to unit stress:

$$v_{C1} = \frac{V_{C1}}{b^2} = \frac{21 \text{ kip}}{(20 \text{ in.})^2} = 53 \text{ psi}$$

expressed as a function of $\sqrt{f_c'}$ (see Chapters 28 and 29)

$$\frac{v_{C1}}{\sqrt{f_c' \text{ psi}}} = 0.8$$

which suggests that the shear stress in the column, if all the assumptions are correct, is moderate. If this result had exceeded 2, it would have demanded careful reanalysis or special transverse reinforcement, even though, as we shall see in Chapters 28 and 29, professional experience is light with structural elements subjected to shear parallel to their diagonal axes.

BARE ESSENTIALS

Shrinkage strain ε_{sh} is not likely to exceed 0.0005. The creep coefficient m is not likely to exceed 3.

If environmental humidity does not vary by more than 15%, two-thirds of the total time-dependent change is likely to occur within three months.

EXERCISES

1. Estimate the order of magnitude of the total shortening of each column in the first floor of Figure 5.9 after one year. Assume that the dimensions of the column cross sections are 22 × 22 in. and $f'_c = 5000$psi.
2. Repeat Example 6.2 for column A1 in Figure 6.9. You may make assumptions similar to those in Example 6.2.

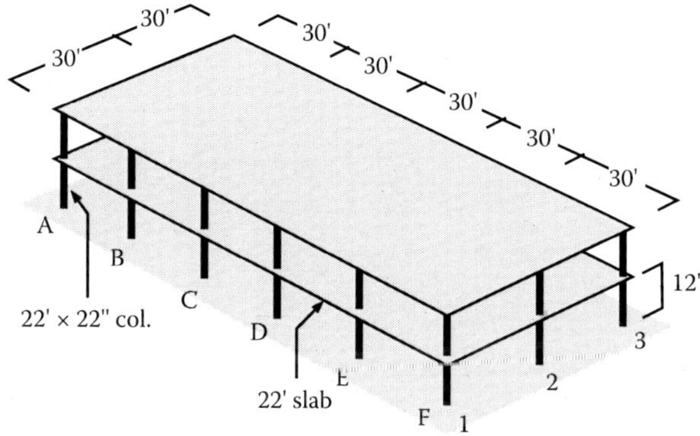

FIGURE 6.9 Exercise 2.

7 Tied Columns

A tied column is a concrete column reinforced with transverse reinforcement called ties (Figure 7.1) in addition to longitudinal reinforcement. The main function of ties, as specified for tied columns, is to hold the reinforcement in place during casting. Ties may also serve to resist shear, confine the concrete, and prevent the bars from buckling if they are proportioned properly for those tasks.

Given their simple and nonstructural function, it would seem that building codes would have no more than a single line of requirement for ties, but this is not so. Many codes stipulate minimum sizes and maximum spacings as well as geometric arrangements for ties. In proportioning a tied column, it is essential to follow the local requirements for ties.

One of the conclusions from the American Concrete Institute (ACI) Column Investigation (Richart and Brown, 1934) that has stood the test of time is the simple relationship that defines the strength of the concrete in a vertically cast column in terms of the cylinder strength, f'_c.

$$f''_c = k_3 \times f'_c \tag{7.1}$$

where f''_c is the compressive strength of concrete in a vertically cast column, k_3 is 0.85, and f'_c is the strength of concrete determined using 6 × 12 in. test cylinders.

There is, indeed, no reason to expect that k_3 should be unity even if the distributions of cement, water, fine aggregate, and coarse aggregate are identical in the column and in the cylinder. The curing rates in structural members of different size would be different, leading to different strengths at different times. On the other hand, so far there has been no way to arrive at the value of k_3 from first principles. It needs to be determined by tests. The value of 0.85 was based on the results of columns tested with and without longitudinal reinforcement in the course of the ACI Column Investigation.[1] Its being smaller than unity was rationalized by recognizing that in a column, considerably taller than the test cylinder, the coarse aggregate tends to "sink" to the lower parts of the column, driving the water to the top. The migration of the water to the top increases the water-to-cement ratio in the upper parts of the column, leading to strength lower than that in the cylinder. It is to be noted that $k_3 = 0.85$ was not the result of a statistical study but of a judgment call. Strictly, its use should be limited to the ranges of the variables included in the ACI Column Investigation. However, the constant $k_3 = 0.85$ has been used universally for reinforced concrete structures, including horizontally cast elements. We shall assume it

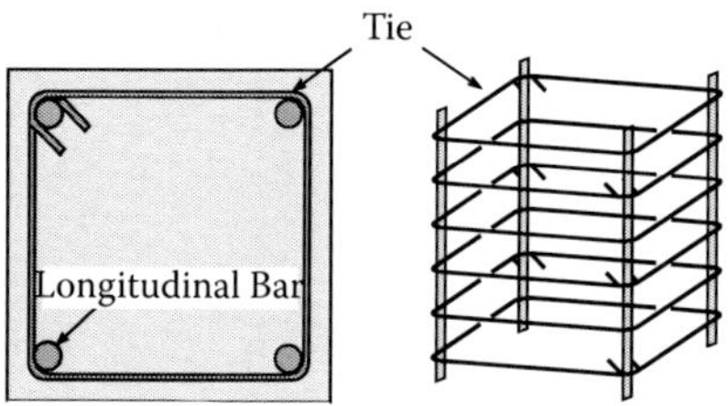

FIGURE 7.1 Tied column.

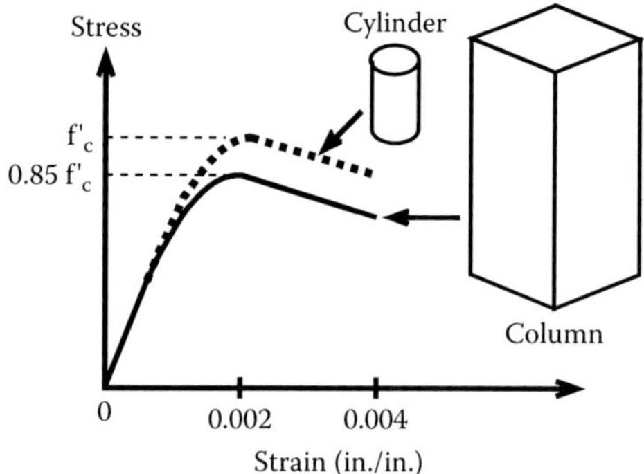

FIGURE 7.2 Idealized stress–strain relationship of concrete cylinder and concrete in a vertically cast column.

to be correct in discussing the strength of the tied column. Figure 7.2 compares idealized stress–strain relationships for a concrete cylinder and for concrete in a column.

If the loading is axial and if the load is one of gravity, that part of the curve beyond maximum unit stress is of little significance. The loaded element would fail at the strain corresponding to reduction in strength unless the load could be reduced at the same rate the strength is reduced with increase in compressive strain.

It may also be self-evident that strains exceeding 0.004 can be measured provided the applied load is also reduced appropriately. In fact, concrete will resist some finite stress, however little, at compressive strains on the order of 0.01 (see Figure 5.3).

The reason we continue the stress–strain diagram beyond that at maximum stress is because it is of significance in relation to flexural response of a section that will be discussed in Chapter 10. Why we stop it at a value of 0.004 has been discussed in Chapter 5.

Nominal yield strains of Grade 60 and 80 bars are

$$\varepsilon_y = \frac{f_y}{E_s} = \frac{60}{29 \times 10^3} \approx 0.0021 \quad \text{(Figure 7.3a)}$$

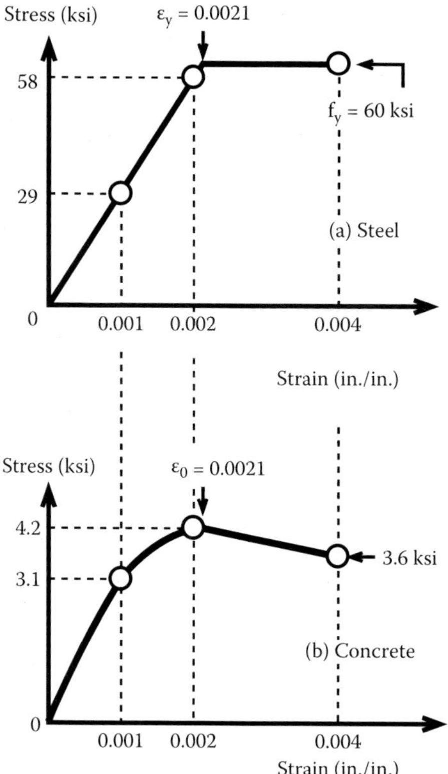

FIGURE 7.3 Stress–strain relationship of Grade 60 bar and 5000 psi concrete.

and

$$\varepsilon_y = \frac{f_y}{E_s} = \frac{80}{29 \times 10^3} \approx 0.0027$$

even though some codes for materials permit values as high as 0.005 for ε_y. From these values and Figure 7.2, it can be inferred that the compressive strength of the concrete and the yield stress of the reinforcement will be reached at approximately the same strain and can be summed to obtain a close estimate of the maximum axial load capacity of a tied column.

$$P_n = \left(A_g - A_s\right) \cdot k_3 \cdot f_c' + A_s \cdot f_y \tag{7.2}$$

where P_n is the axial load capacity of a tied column, k_3 is 0.85, A_g is the total (gross) area of the column cross section, A_s is the cross-sectional area of longitudinal reinforcement, f_c' is the compressive strength of concrete determined by 6 × 12 in. test cylinders, and f_y is the yield stress of longitudinal reinforcement.

Equation 7.2 is also used in a dimensionless form:

$$\frac{P_n}{A_g \cdot f_c'} = k_3 \cdot (1-\rho) + \rho \frac{f_y}{f_c'} \tag{7.3}$$

where ρ is the longitudinal reinforcement ratio, A_s/A_g.

Equation 7.3 enables a quick estimate of the relative contribution of reinforcement to the capacity of the concrete. For example, if the reinforcement ratio ρ is 1%, the first term on the right-hand side becomes $0.99{*}k_3$ and may be taken as 0.85. If the ratio of the yield stress to the cylinder strength, f_y/f_c', is 10, the second term becomes 0.1, indicating that, for the column considered, concrete is the main source of strength in axial compression.

EXAMPLE 7.1

A tied column has a clear height of 10 ft and a cross section measuring 20 in.^{2}. Its material properties are listed below:

Longitudinal reinforcement: Eight #7 bars, Grade 60.
Young's modulus for reinforcement: E_s = 29,000,000 psi.
Concrete strength: f_c' = 5000 psi.
Young's modulus for concrete: E_c = 4,000,000 psi.
Transverse reinforcement: Pairs of #3 ties at 12 in.

Our goal is to plot the relationship between the applied axial force, P, and displacement at the top of the column, δ, at unit strains of 0.001, 0.002, and 0.004.

The first step is to express the amount of longitudinal reinforcement as a ratio of the column cross-sectional area.

The cross-sectional area of each #7 bar is A_{s7} = 0.6 in.^{2}.
Total cross-sectional area of longitudinal reinforcement $A_s = 8A_{s7}$ = 4.8 in.^{2}.
Column width b = 20 in.
Gross cross-sectional area of column $A_g = b^2$ = 400 in.^{2}.
Reinforcement ratio $\rho = \frac{A_s}{A_g} = 1.2\%$.

The stress–strain relationship of reinforcing steel is shown in Figure 7.3a.

In accordance with the expression by Hognestad (Chapter 5), the strain ε_0 at maximum stress of concrete is

$$\varepsilon_0 = \frac{2 \times 0.85\, f_c'}{E_c} = 0.0021$$

Concrete stress at a strain ε_1 = 0.001:

$$f_{c1} = 0.85\, f_c' \left[2\frac{\varepsilon_1}{\varepsilon_o} - \left(\frac{\varepsilon_1}{\varepsilon_0}\right)^2 \right] = 3100 \text{ psi (Figure 7.3b)}$$

Concrete stress at a strain $\varepsilon_2 = 0.002$:

$$f_{c2} = 0.85\,f_c'\left[2\frac{\varepsilon_2}{\varepsilon_o} - \left(\frac{\varepsilon_2}{\varepsilon_0}\right)^2\right] = 4200 \text{ psi}$$

Concrete stress at a strain $\varepsilon_3 = 0.004$:

$$f_c = 0.85\,f_c'\left[1 - \frac{\varepsilon_3 - \varepsilon_o}{0.004 - \varepsilon_o}\cdot 0.15\right] = 3600 \text{ psi}$$

Reinforcement stress at $\varepsilon_1 = 0.001$:

$$f_{s1} = \text{Young's modulus} \times \text{strain} = 29{,}000 \text{ ksi} \times 0.001 = 29 \text{ ksi}$$

Reinforcement stress at $\varepsilon_2 = 0.002$:

$$f_{s2} = \text{Young's modulus} \times \text{strain} = 29{,}000 \text{ ksi} \times 0.002 = 58 \text{ ksi}$$

Reinforcement stress at $\varepsilon_3 = 0.004$:

$$f_{s3} = \text{Yield stress} = 60 \text{ ksi}$$

Axial force developed at a strain of 0.001 (Figure 7.4) is

$$\begin{aligned} P_{0.001} &= \text{Force in concrete} + \text{Force in reinforcement} \\ &= \left(A_g - A_s\right) \times f_{c1} + A_s f_{s1} = (400 - 4.8)\text{ in.}^2 \times 3.1 \text{ ksi} + 4.8 \text{ in.}^2 \times 29 \text{ ksi} \\ &= 1220 \text{ kip} + 140 \text{ kip} = 1360 \text{ kip} \end{aligned}$$

Axial force developed at a strain of 0.002 is

$$\begin{aligned} P_{0.002} &= \left(A_g - A_s\right) \times f_{c2} + A_s f_{s2} = (400 - 4.8)\text{ in.}^2 \times 4.2 \text{ ksi} + 4.8 \text{ in.}^2 \times 48 \text{ ksi} \\ &= 1660 \text{ kip} + 280 \text{ kip} = 1940 \text{ kip} \end{aligned}$$

Axial force developed at a strain of 0.004 is

$$\begin{aligned} P_{0.004} &= \left(A_g - A_s\right) \times f_{c3} + A_s f_{s3} = (400 - 4.8)\text{ in.}^2 \times 3.6 \text{ ksi} + 4.8 \text{ in.}^2 \times 60 \text{ ksi} \\ &= \sim 1400 \text{ kip} + 290 \text{ kip} = 1690 \text{ kip} \end{aligned}$$

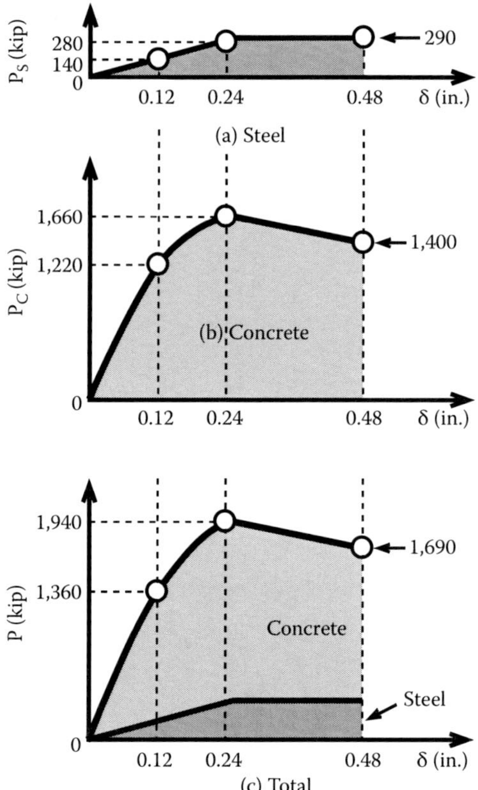

FIGURE 7.4 Axial force vs. displacement.

Shortening of column at a strain of 0.001 is

$$\delta_{0.001} = \text{height} \times \text{strain} = 10 \text{ ft} \times 0.001 = 120 \text{ in.} \times 0.001 = 0.12 \text{ in.}$$

Shortening of column at a strain of 0.002 is

$$\delta_{0.002} = \text{height} \times \text{strain} = 10 \text{ ft} \times 0.002 = 120 \text{ in.} \times 0.002 = 0.24 \text{ in.}$$

Shortening of column at a strain of 0.004 is

$$\delta_{0.004} = \text{height} \times \text{strain} = 10 \text{ ft} \times 0.004 = 120 \text{ in.} \times 0.004 = 0.48 \text{ in.}$$

Even though our calculations have led to shortening results printed to 1/100 of an inch, we realize that 1/100 in. represents unreasonable precision for a 10 ft reinforced concrete column. When we use them in a professional environment we call them 1/8, 1/4, and 1/2 in.

Summary of Results

Unit Strain (Compression)	Axial Load (kip)	Shortening (in.)
0	0	0
0.001	1350	1/8
0.002	1940	1/4
0.004	1690	1/2

7.1 DESIGN STRENGTH OF AXIALLY LOADED SHORT COLUMNS

The strength for an axially loaded short* column calculated using Equation 7.2 refers to laboratory conditions, and for those conditions it is quite good provided the longitudinal reinforcement has a yield stress not exceeding 80,000 psi and the loading to failure is completed within a period of hours. Under construction conditions, the expression needs to be modified for two dominant reasons. One is the likelihood of accidental eccentricities, and the other is that of errors in dimensions and material strength. The required modification is a matter of professional judgment and will vary from country to country and over time, depending on practice and concepts of acceptable risk. In Example 7.2, Equation 7.2 is modified by 1/2. Considering that the load factor should not be below 1.4, the overall factor would not be below 2.8. If the sustained load is below half the short-time capacity of the column, it is very unlikely that it will reduce the long-time strength of the column (Chapter 6).

EXAMPLE 7.2

Determine the limiting load to be used in design for the column section described in Example 7.1.

SOLUTION

$$P_{design} = \frac{1}{2}\cdot\left[0.85\cdot f_c'\cdot\left(A_g - A_s\right) + f_y\cdot A_s\right]$$

$$= \frac{1}{2}\cdot\left[0.85\cdot 5\text{ ksi}\cdot\left(400\text{ in.}^2 - 4.8\text{ in.}^2\right) + 60\text{ ksi}\cdot 4.8\text{ in.}^2\right] = 980\text{ kip}$$

* A column that is not susceptible to buckling, i.e., a column with restrained ends and a minimum cross-sectional dimension exceeding one-sixth of its clear height.

BARE ESSENTIALS

The compressive strength of the concrete in a tied column is

$$f_c'' = k_3 f_c'$$

where $k_3 \approx 0.85$.

The strength of a tied column can be estimated as

$$P_n = (A_g - A_s) k_3 f_c' + A_s f_y$$

As long as the longitudinal reinforcement does not exceed 2% and its yield stress is 80 ksi or lower, the axial strength of the column can be estimated for preliminary design as the product of the gross section area and the cylinder strength modified by 0.85.

If the sustained unit stress in an axially loaded column is ~1500 psi on initial loading, the corresponding stress in the reinforcement may be ~12,000 psi. With time, the stress in the reinforcement may reach a value as high as 50,000 psi. The reinforcement may even yield if the reinforcement ratio is below 1%. But that condition does not reduce the nominal strength of the column.

EXERCISES

1. Repeat Example 7.2 assuming that we use #9 bars. You will note that the contribution of steel to the design axial strength is still smaller than that of concrete.
2. Select cross-sectional dimensions and reinforcement for the first story of the interior column of the 5-story office building shown in Figure 5.9.

   Assume that concrete of nominal strength = 5000 psi and reinforcing bars with f_y = 60 ksi are to be used.

   Assume that the dead and live loads per unit area are 200 psf (including self weight) and 30 psf, respectively. Consider load combinations: (1.4 DL) and (1.2 DL + 1.6 LL). Draw a cross section of the column with the selected properties (dimensions and reinforcement).
3. Repeat Example 7.2 using the gross area for the contribution of the concrete to axial strength (ignoring the reduction to obtain the net concrete area, ($A_g - A_s$)). What is the difference in strength so determined compared with the result obtained for Example 7.2?

8 Axial Strength of Laterally Confined Concrete

To understand the effects of the factors defining the axial strength of laterally confined concrete, we start by examining a sample result from a series of tests of columns under axial compression by Richart et al. (1929). The column measured 10 in. around by 40 in. long. It had no longitudinal reinforcement, but had spiral reinforcement (Figure 8.1a) or helical reinforcement* on the outer surface of the test column. The spiral reinforcement had a diameter of 0.19 in., and its spacing was 1 in. The cylinder strength of the concrete in the column was 2640 psi. The spiral reinforcement developed 61 ksi in tension at a strain of 0.005. Figure 8.1 shows the variation of the measured axial and radial strains in the concrete core with increase in axial stress.

The solid curve in Figure 8.1b refers to the longitudinal compressive strain (plotted as positive strain). The broken curve refers to the radial strain (plotted as negative strain). It should be emphasized that the radial unit strain is the expansion of the column diameter divided by the initial diameter. A radial strain, so measured, of more than 0.0002 is very likely an indication of the initiation of longitudinal cracks in the concrete, and the reported value is not material strain, but simply the expansion of the diameter divided by the original diameter. At a radial strain approaching 0.0003, a perceptible change in the measured radial strain rate is observed, confirming initiation of internal cracking.

The relationship between the compressive stress and longitudinal strain is similar to the relationship described in Chapter 5 for a standard test cylinder. The measured curve starts out nearly linear. Its slope decreases as the load is increased. The important observation is that the compressive strain is accompanied by a radial tensile strain. We understand the relationship in the initial part of the curve by remembering what we know of Poisson's ratio. An axially loaded material expands laterally, the rate of expansion depending on its density. But what occurs in concrete after the apparent radial strain exceeds 0.0002 and vertical cracks (Figure 5.3c) form requires further thinking. It becomes difficult to rationalize the relationship between the radial and longitudinal strains in terms of Poisson's ratio, which relates to a continuous material.

We can interpret the observed phenomenon by considering the idealized two-dimensional truss model shown in Figure 8.2. The model comprises four rods with frictionless hinges at intersection points. A horizontal spring joins two of the hinges. A vertical load is applied at the top hinge. A reaction is provided at the bottom hinge.

* In engineering use, helical reinforcement has been designated incorrectly as spiral reinforcement. In the following text, we shall conform to engineering use.

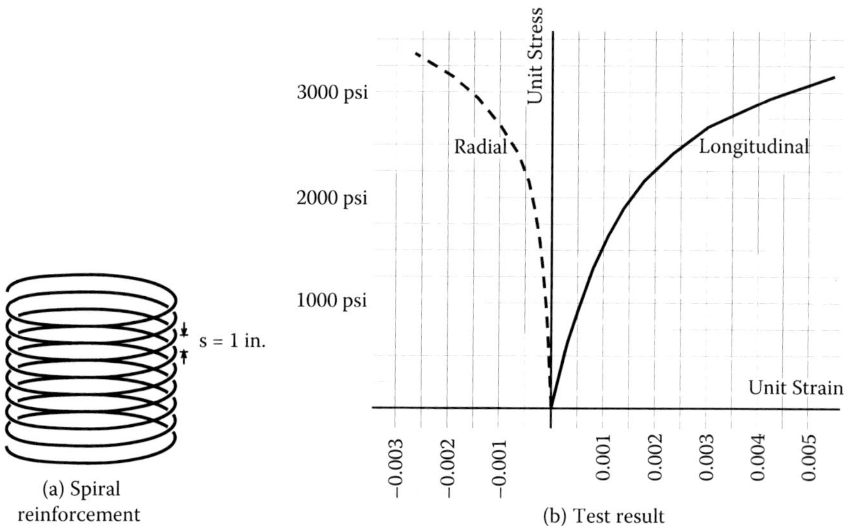

FIGURE 8.1 Stress–strain relationships for confined concrete.

By visualizing the deformation of the hinged truss, we sense that the vertical load will cause compression in the rods and tension in the spring. We stretch our imagination to assume that the rods and the spring are all made of concrete. We know that concrete is strong in compression and weak in tension. The ratio is approximately 10 to 1. We conclude that the capability of the truss to resist the vertical load will be controlled by the tensile strength of the spring.

The simple model in Figure 8.2 also helps us understand the unit strain measurements made in the column tests. As the vertical load is increased, there will be a shortening in the height of the truss (compressive strain) and an increase in the width of the truss (tensile strain.)

If the vertical load is increased to a level at which the strain in the spring reaches its limit, the spring will snap and the truss will fail. That is one way to understand the failure in compression of a test cylinder of plain concrete.

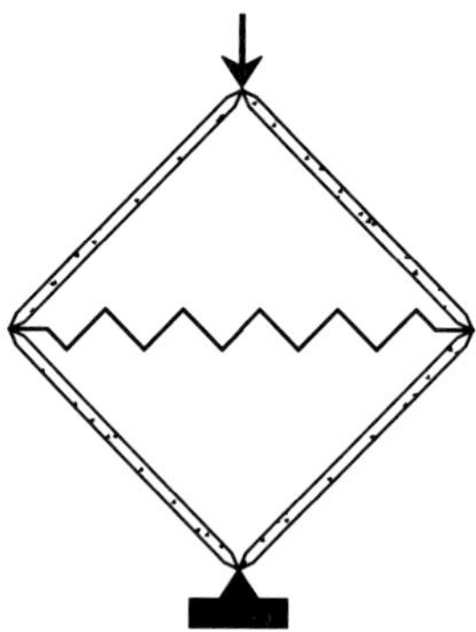

FIGURE 8.2 Truss.

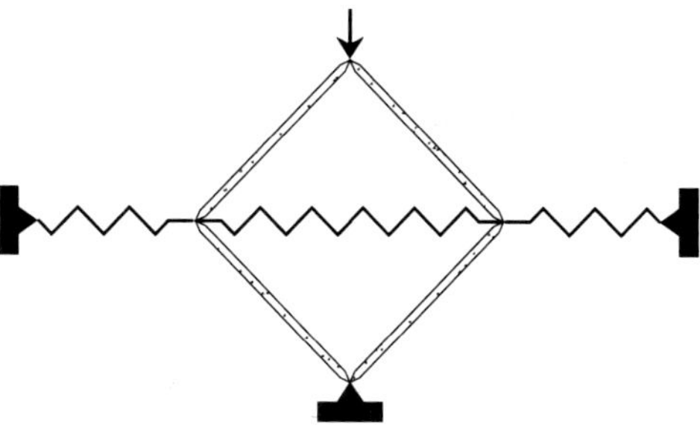

FIGURE 8.3 Truss with external springs.

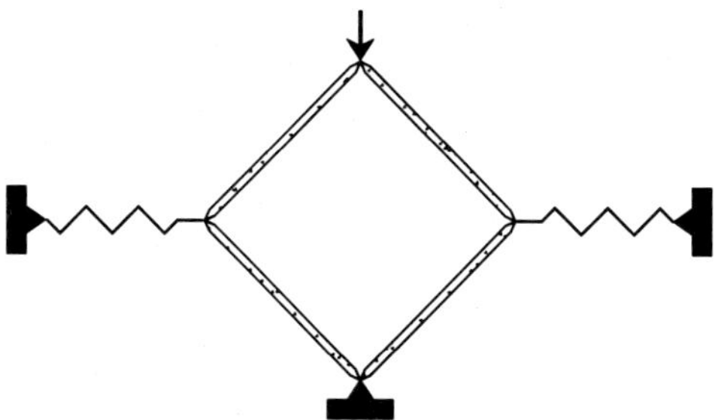

FIGURE 8.4 Truss after rupture of internal spring.

Figure 8.3 shows a different arrangement of the same truss model. In this case, there are external springs that continue on to supports that provide horizontal reactions. As the vertical load is applied, the internal spring develops tension, while the external springs develop compression. At a certain level of the applied load, the internal spring (weak in tension) will snap, but the truss will not fail because the external springs (strong in compression) will supply the required force to maintain equilibrium. The truss will continue to resist increased levels of the vertical load because the external concrete springs will be in compression. In that state, the truss model can be visualized as illustrated in Figure 8.4.

Richart and his associates (1929) desired to quantify the rate at which the axial strength increased for a given lateral (confining) stress. They considered an axially loaded column with a circular section as shown in Figure 8.5 with no longitudinal reinforcement, but with, ideally, circular hoops at a small spacing along the height of the column. To relate the force in the hoops to the lateral stress on the concrete within the hoops, they used a brilliantly simple approach.

The spacing of the hoops in the column illustrated (Figure 8.5a) is defined by s. The diameter of the column section is d_c. Consider a segment of the column of length s (Figure 8.5b). Cut a vertical section along any diameter (Figures 8.5c and 8.6). The cross-sectional area of the vertical section A_c is

$$A_{ct} = d_c s \tag{8.1}$$

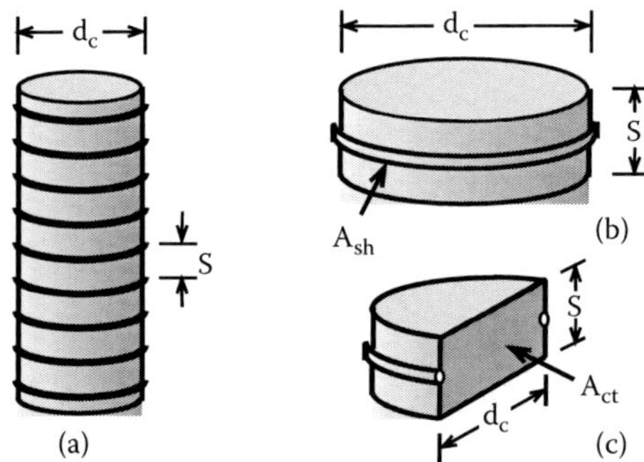

FIGURE 8.5 Column with hoops.

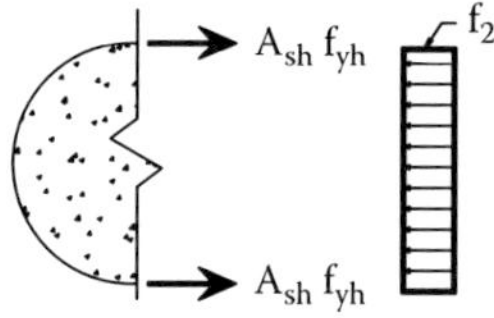

FIGURE 8.6 Transverse forces acting on the concrete core.

If it is assumed that the unit stress in the hoop is its yield stress, f_{yh}, and if it is also assumed that the lateral stress on the concrete across the diameter is uniform (Figure 8.6), from the equilibrium of forces acting on the segment:

Total force exerted by hoop = Force on concrete section

$$2A_{sh}f_{yh} = f_2 d_c s \tag{8.2}$$

where A_{sh} is the cross-sectional area of the hoop bar, f_{yh} is the yield stress of the hoop bar, and f_2 is the lateral unit stress on confined concrete.

For convenience, we define ρ_h (the ratio of volume of the hoop to the volume of the confined segment of the column with diameter d_c and thickness s):

$$\rho_h = \frac{A_{sh} \cdot \pi \cdot d_c}{\dfrac{\pi \cdot d_c^2}{4} \cdot s}$$

or

$$\rho_h = \frac{4A_{sh}}{d_c s} \tag{8.3}$$

Substituting the definition of the volumetric ratio of hoops in the equilibrium statement,

$$2 \cdot \frac{d_c \cdot s \cdot \rho_h}{4} \cdot f_{yh} = f_2 \cdot d_c \cdot s \tag{8.4}$$

Solving for f_2,

$$f_2 = \frac{f_{yh} \cdot \rho_h}{2} \tag{8.5}$$

Equation 8.5 provides a simple relationship for the unit confining stress imposed by the amount and yield stress of the hoop bars. The same expression can be used for spiral reinforcement or helical reinforcement with uniform spacing s.

Richart and coworkers tested columns with spiral reinforcement under axial load. These test specimens had no longitudinal reinforcement and no shell (no concrete cover on the spiral reinforcement). Their results are summarized in Figure 8.7. They found that, for the range of concrete properties tested, the relationship between the maximum unit axial stress, f_1, in the concrete and the calculated maximum confining stress, f_2, was defined closely by the expression

$$f_1 = f_c'' + 4.1\ f_2 \tag{8.6}$$

where f_c'' is the strength of the concrete in the column, a strength that Richart observed to be 85% of the 6 × 12 in. cylinder strength. In this book, we replace 4.1 with 4 for simplicity. Figure 8.8 shows a schematic diagram of the stress–strain relationship of plain concrete and confined concrete.

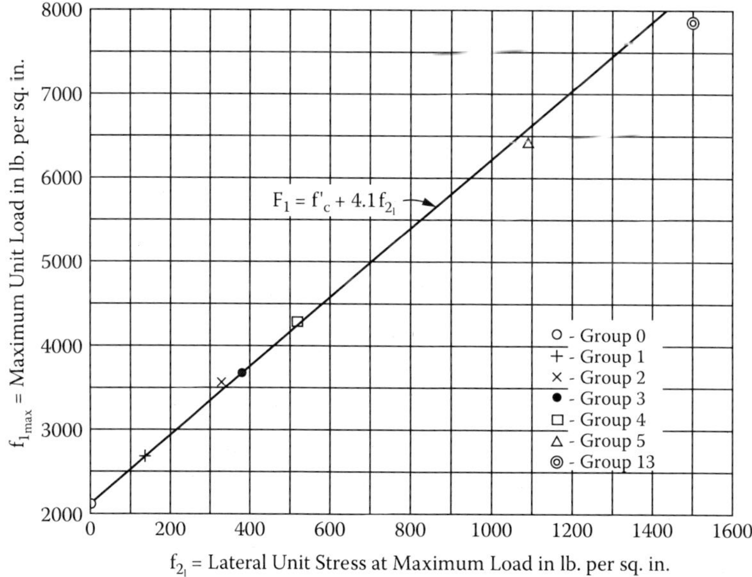

FIGURE 8.7 Maximum axial stress vs. lateral unit stress.

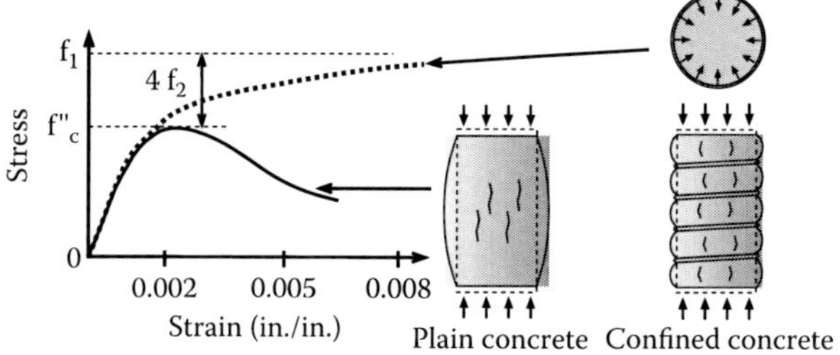

FIGURE 8.8 Stress–strain relationship of plain concrete and confined concrete.

EXAMPLE 8.1

Figure 8.9 shows the cross section of a column with hoops at a uniform spacing of 2 in. The hoops have been cut from standard Grade 60 #3 bars, and the nominal strength of the concrete is 6000 psi. Compute the maximum stress that can be reached in the concrete core.

SOLUTION

The volumetric ratio of the hoops is

$$\rho_h = \frac{4 \times 0.11\,\text{in.}^2}{16\,\text{in.} \times 2\,\text{in.}} = 0.014$$

The lateral unit stress on confined concrete is

$$f_2 = \frac{60{,}000\,\text{psi} \times 0.014}{2} = 420\,\text{psi}$$

The maximum stress that can be developed in the core concrete is

$$f_1 = 0.85 \times 6000\,\text{psi} + 4 \times 420\,\text{psi} = 5100\,\text{psi} + 1700\,\text{psi} = 6800\,\text{psi}$$

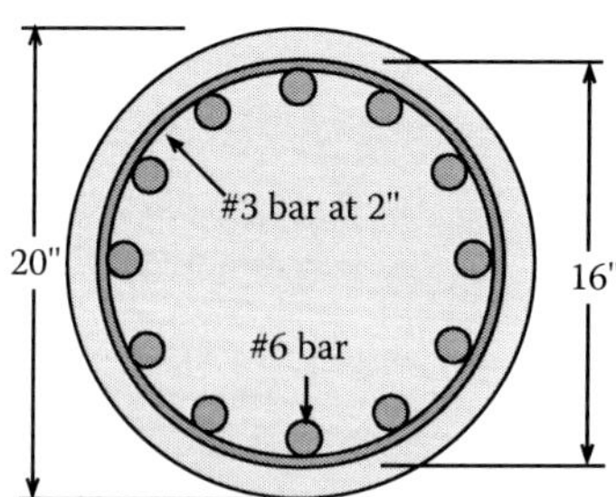

FIGURE 8.9 Example 8.1.

BARE ESSENTIALS

- The concrete core of an axially loaded column confined by spiral (helical) reinforcement will have increased strength and deformability.
- For normal weight aggregate concrete, the increase in strength of confined concrete is approximately $4f_2$, where f_2 is half of the product of the volumetric ratio of the spiral reinforcement and its effective yield stress for $f_2 \leq 2000$ psi.

EXERCISES

1. Repeat Example 8.1 assuming the spacing of the hoops to be 1 in.
2. Repeat Example 8.1 assuming the spacing of the hoops to be 3 in.
3. Draw the envelope of the Mohr circles defined by the unit stresses f_1 and f_2 obtained in Exercises 1 and 2 and in Example 8.1. Find its intercept with the vertical axis (representing unit shear stress) and its slope. Note: f_1 and f_2 represent principal stresses.

9 Spiral Columns

The spiral column is not literally a column having the shape of a spiral, a two-dimensional geometric form (Figure 9.1). It is a concrete column reinforced with helical (Figure 9.2) in addition to longitudinal reinforcement. It has a circular core confined by helical reinforcement wound at a spacing that is typically not smaller than 1 in. and that rarely exceeds 3 in. In engineering use, helical reinforcement has been called spiral reinforcement, and the column with spiral reinforcement is called a spiral column. Representative spiral column sections are shown in Figure 9.3.

Spiral reinforcement is protected by a concrete cover (shell). If the gross section of the column (core plus shell) is rectangular, additional longitudinal reinforcement may be placed in the corner regions if desired.

In this section we treat the proportioning of axially loaded spiral columns, a rather simple design task. No matter how simple, design is an act that demands the use of judgment as well as science. Judgment involves the vast experience of the profession acquired over centuries and the discrimination of the individual engineer. There is seldom a unique solution to an engineering problem. Design requires creative thinking at all levels. The thought process for selecting the size, shape, and reinforcement of a simple axially loaded column is akin to that in, say, the selection of the dimensions of a suspension bridge. It requires knowledge of geometry and mechanics as well as a sense of proportion. In addition to strength, serviceability, and durability, the designer needs to be concerned with constructability. The reinforcement arrangement selected has to facilitate concrete casting and connectivity with reinforcement in elements continuous with the column.

A spiral column differs from a tied column in that part of the strength of the spiral column comes from confined concrete, from the increase in axial (longitudinal) compressive strength of the concrete made possible by the spiral reinforcement. Spiral reinforcement also increases shear strength and toughness.

The relationship of the materials in the spiral column to its axial strength is similar to the relationship of the properties of an organic compound to its individual elements. The strength of the spiral column is a complex combination of three components: (1) the longitudinal reinforcement, (2) the concrete in the shell, and (3) the concrete in the core. These three components contribute different fractions of the total resistance at different levels of compressive strain.

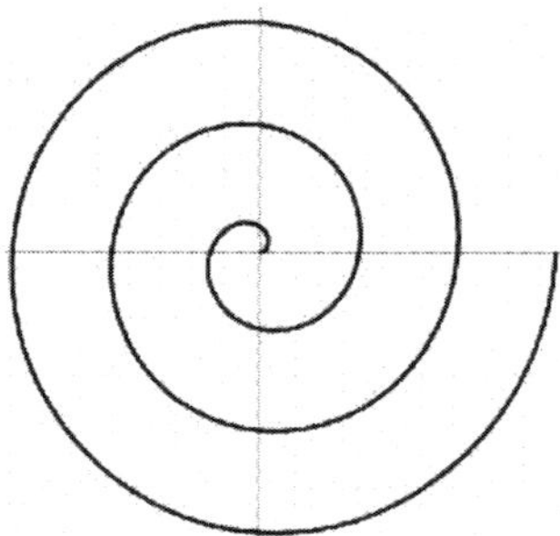

FIGURE 9.1 Spiral.

FIGURE 9.2 Helix.

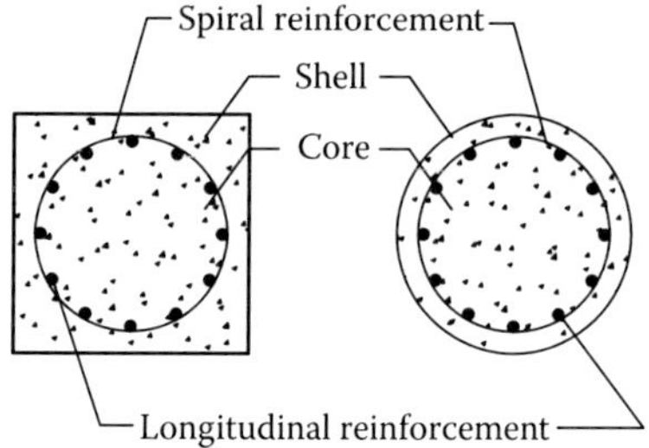

FIGURE 9.3 Cross sections of spiral columns.

9.1 STRENGTH COMPONENTS OF A SPIRAL COLUMN IN AXIAL COMPRESSION

In the process of proportioning a spiral column, it is assumed that the longitudinal reinforcement reaches its yield stress. Different upper bounds to the yield stress that can be developed have been set by different codes in the range of 60 to 80 ksi.

Accordingly, the contribution of the longitudinal reinforcement is

$$P_s = A_s \cdot f_y \tag{9.1}$$

where A_s is the total cross-sectional area of longitudinal reinforcement and f_y is the yield stress of longitudinal reinforcement.

The limit defined by Equation 9.1 ignores the possible increase in steel stress related to strain hardening of the types of steel used for reinforcement.

Assuming that the maximum stress to be reached in the shell concrete is limited to 85% of cylinder strength, the limiting contribution of the concrete in the shell is

$$P_{shell} = (A_g - A_c) \cdot 0.85 \cdot f'_c \tag{9.2}$$

where A_g is the cross-sectional area of the entire section, A_c is the cross-sectional area of the column core confined by the spiral reinforcement, and f'_c is the strength of the standard test cylinder (6 × 12 in.).

The variation of the resistance of the shell concrete with compressive stress is assumed to be identical to that for a test cylinder. The strain at which the maximum stress is attained depends on several factors, such as strength and rate of loading. In this section we shall ignore the time-dependent volume changes and strain rate effects in concrete. We shall also assume that, for normal weight aggregate concretes with cylinder strengths not exceeding 6000 psi, the maximum stress is reached at a strain ε_o close to 0.002 (Chapter 5). Beyond strain ε_o, the resistance of the concrete decreases as suggested in Figure 5.8.

As discussed in Chapter 8, the maximum resistance contributed by the concrete confined by spiral reinforcement is defined approximately by the expression

$$P_{cc} - A_c \cdot (0.85 \cdot f'_c + 2 \cdot \rho_h \cdot f_{yh}) \tag{9.3}$$

where A_c is the cross-sectional area of column core confined by spiral reinforcement, f'_c is the strength of the standard test cylinder (6 × 12 in.), ρ_h is the volumetric ratio of spiral reinforcement, and f_{yh} is the yield stress of spiral reinforcement.

The additional strength of the core concrete, made possible by the transverse (spiral) reinforcement, comes at the expense of axial strain. An example (Richart et al., 1929) is shown in Figure 9.4. The data presented are from the test of a column with a diameter of 10 in. and a length of 40 in. loaded axially to failure during a short time. The test column had no longitudinal reinforcement and no shell. The volumetric ratio of the transverse reinforcement (see Chapter 8 for definition) was 4.4%. It is seen in Figure 9.4 that the core, with 2700 psi concrete, developed a resistance exceeding 6000 psi. But this stress was achieved at an axial strain of 2.4%. If that particular column had had a shell, the contribution of the shell would have been lost by the time the core developed 4000 psi.

The axial deformation of a particular spiral column with a shell is represented in Figure 9.5. At an axial unit strain of 0.002 the strength of the unconfined concrete section, including the core and the shell, is developed along with the yield stress of the longitudinal reinforcement. In effect, that is the strength of the tied column:

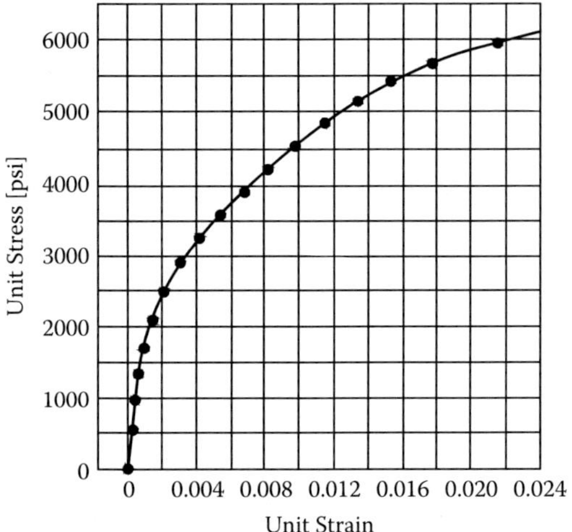

FIGURE 9.4 Stress–strain relationship measured for a spiral column.

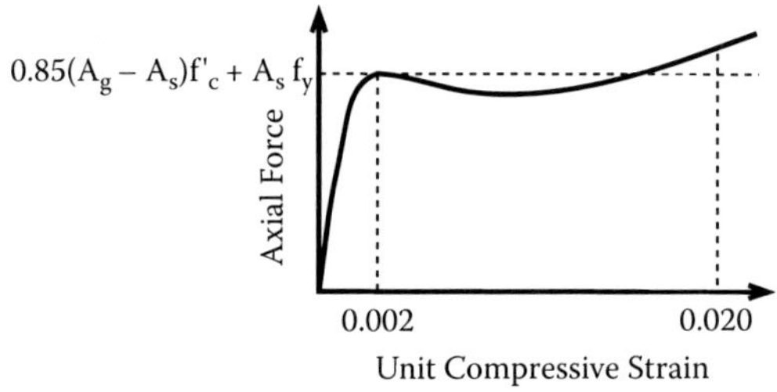

FIGURE 9.5 Idealized axial force–strain response of a spiral column.

$$P_n = 0.85 \cdot (A_g - A_s)\, f_c' + A_s \cdot f_y \tag{9.4}$$

where A_g is the total (gross) area of the cross section, f_c' is the compressive strength of the concrete determined by tests of 6 × 12 in. cylinders, A_s is the total area of longitudinal reinforcement, and f_y is the yield stress of reinforcement (60 ksi in this case).

With increase in strain, the unconfined concrete in the shell starts deteriorating, while the core concrete gains strength. In the case shown, the strength developed at an axial unit strain of 0.02 exceeds that developed at 0.002.

In building codes used currently, the strength of the spiral column (the second peak in Figure 9.5) is assumed to be equal to the strength of the tied column (the first peak in Figure 9.5). However, the strength reduction factor used to determine the design strength of the spiral column is higher than that used for a tied column. This approach is based on the majority opinion of Committee 105 of the American Concrete Institute (ACI Committee 105, 1933). It was arrived at after heated debate within the committee that forced the committee to issue a majority and a minority report. The approaches adopted by the two groups are worth summarizing because it reveals how codes of practice are made and the subtleties of the use of engineering judgment.

Briefly, the decision of the majority of Committee 105 was to define the strength of the spiral column as that of the equivalent tied column, but to permit a lower factor of safety for spiral columns. To justify the reduction in the factor of safety, a minimum requirement of spiral reinforcement was set. The minimum was based on the condition that the second peak in Figure 9.5 would not be below the first peak, or that the increase in strength of the core made possible by the spiral reinforcement would compensate the loss of the shell.

From that condition, the reasonable amount of minimum volumetric ratio of spiral reinforcement was judged to be

$$\rho_h = 0.45 \cdot \left(\frac{A_g}{A_c} - 1 \right) \cdot \frac{f'_c}{f_{yh}} \tag{9.5}$$

where ρ_h is the volumetric ratio of spiral reinforcement, A_g is the total cross-sectional area of the column section including the shell and the core, A_c is the cross-sectional area of the column core confined by spiral reinforcement, f'_c is the strength of the standard test cylinder (6 by 12 in.), and f_{yh} is the yield stress of spiral reinforcement.

The student should note that the derivation of Equation 9.5 refers to axial loading, but the conditions in actual structures may involve bending combined with axial load. It should be clear that the amount indicated above is a result of the judgment of some of the committee members and does not represent a conclusion based on mechanics, even though it appears to be so.

The minority group of Committee 105 did not agree that such a simplification should be made. Instead, they proposed that both peaks (in Figure 9.5) should be computed with a limit on the deformation for the second maximum. They reasoned that such an approach was closer to the observed phenomenon. And they were indeed correct, but the majority argued that the complications were not worth the benefit achieved, and that because the result was not going to be accurate anyway, it was better to be wrong the easy way.

The student is encouraged to read the report of Committee 105 because it contains, in a few pages, a very instructive example of the role of judgment in engineering design.

EXAMPLE 9.1

Select reinforcement and dimensions for a spiral column to support a load of 600 kip (that compares well with weight supported by an interior column in the first story of a 7- to 8-story reinforced concrete building with 20 ft spans).

The column, with a circular cross section, is to have a clear height of 10 ft. It is to be built using 5000 psi concrete and Grade 60 longitudinal reinforcement. The overall factor of safety (load factor divided by strength reduction factor) to be used is 2.5 (which is less than what would be used for tied columns).

SOLUTION

The first move is to guess at the size of the column. From the experience of designers in Chicago, we decide to use approximately 1% longitudinal reinforcement. From the proportions of classical columns, we guess the diameter of the gross section to be approximately

$$\text{Diameter} \approx \frac{10\text{ ft}}{7} = 17\text{ in.}$$

We choose to use an even number: 18 in. We shall first proportion the section as a tied column, and then add the required spiral reinforcement.

The load demand is

$$P_u = 2.5 \cdot 600\text{ kip} = 1500\text{ kip}$$

With the reinforcement ratio determined and the concrete strength specified, the nominal strength of the cross section, evaluated as a tied column, is defined as

$$P_n = A_g\,[0.85f'_c\,(1 - \rho) + \rho\, f_y]$$

Equating the demand to the resistance and rearranging the terms to determine A_g,

$$A_g = \frac{1500\text{ kip}}{(1-\rho)\cdot 0.85f'_c + \rho \cdot f_y} = 312\text{ in.}^2$$

$$\text{Column diameter required: } \sqrt{\frac{4}{\pi}\cdot 312\text{ in.}^2} = 20\text{ in.}$$

The result of our calculation is within ~10% of our initial guess. In this case, we feel confident that the result we have obtained is good. Nevertheless we shall check our arithmetic after we choose the size and number of reinforcing bars. The total cross-sectional area required is

$$A_s = 0.01\frac{\pi}{4}(20\text{ in.})^2 = 3.1\text{ in.}^2$$

We can select six #7 bars providing

$$A_s = 6 \cdot 0.6\text{ in.}^2 = 3.6\text{ in.}^2$$

or we can select four #8 bars providing

$$A_s = 4 \cdot 0.79\text{ in.}^2 = 3.2\text{ in.}^2$$

The second choice appears to be more economical, but we prefer the first choice considering that the spiral reinforcement will be supported more effectively.

We can now check our arithmetic using a simple procedure, by determining the strength components provided by the concrete and by the reinforcement. The strength component provided by the concrete is

$$\left[\frac{\pi}{4}\cdot(20\text{ in.})^2 - 3.6\text{ in.}^2\right]\cdot 0.85\cdot 5\text{ ksi} = 1320\text{ kip}$$

The component attributed to the steel is

$$3.6\text{ in.}^2\cdot 60\text{ ksi} \approx 215\text{ kip}$$

The sum of the contributions of concrete and steel is above 1500 kip. The arithmetic used must have been correct.

Because we have selected an overall factor of safety that was based on use of a spiral column, we select the required amount of transverse reinforcement from Equation 9.5. Using a core with a diameter of $d_c = 16$ in. (to leave room for cover) and spiral reinforcement with a yield stress of 75 ksi,

$$\rho_h = 0.45\cdot\left(\frac{\frac{\pi}{4}(20\text{ in.})^2}{\frac{\pi}{4}(16\text{ in.})^2} - 1\right)\cdot\frac{5\text{ ksi}}{75\text{ ksi}} = 1.7\%$$

We chose a spacing of 1.5 in. for the spiral reinforcement. The required cross-sectional area is

$$A_{sh} = \frac{d_c\cdot s\cdot\rho_h}{4} = \frac{16\text{ in.}\cdot 1.5\text{ in.}\cdot 0.017}{4} = 0.10\text{ in.}^2$$

We call for a #3 Grade 75 spiral at a spacing of 1.5 in. (for a volumetric ratio $\rho_h = 0.018$).

The last step in the process is determining the strength of the core. The main function of this step is to check the arithmetic. The confining pressure that may be exerted by the spiral reinforcement is

$$f_2 = \frac{\rho_h\cdot 75\text{ ksi}}{2} = 675\text{ psi}$$

The axial unit strength of the concrete core is

$$f_1 = 0.85\cdot f_c' + 4\cdot f_2 = 7.0\times 10^3\text{ psi}$$

The resistance of concrete and steel confined by the spiral reinforcement is

$$P_c = f_1\cdot\left[\frac{\pi}{4}(16\text{ in.})^2 - 3.6\text{ in.}^2\right] + 60\text{ ksi}\cdot 3.6\text{ in.}^2 = 1.6\times 10^3\text{ kip}$$

The concrete core is deemed to have sufficient strength.

BARE ESSENTIALS

The nominal strength of a spiral column is assumed to be equal to the nominal strength of a tied column:

$$P_n = 0.85 \cdot (A_g - A_s)\, f'_c + A_s \cdot f_y$$

However, the strength reduction factor used to determine the design strength of the spiral column is higher than that used for a tied column as long as the volumetric ratio of spiral reinforcement ρ_h exceeds the minimum defined by the expression

$$\rho_h = 0.45 \cdot \left(\frac{A_g}{A_c} - 1 \right) \cdot \frac{f'_c}{f_{yh}}$$

EXERCISES

1. Repeat Example 9.1 for a spiral column to support a load of 1200 kip. Use 6000 psi concrete and Grade 60 reinforcement.
2. A spiral column measures 24 in.^{2}. The total area of longitudinal reinforcement provided is 9 in.^{2}. The nominal yield stress of the longitudinal reinforcement is 60 ksi. The spiral reinforcement is provided by #3 bars with a yield stress of 80 ksi. The diameter of the concrete region confined by the spiral reinforcement is 20 in.

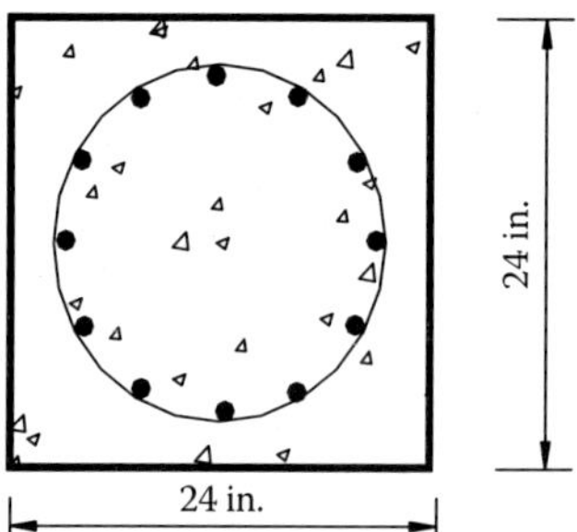

Compressive strength of the normal weight aggregate concrete in the column is 4000 psi. The height of the column is 14 ft.

a. If the tributary area for the column is 20 × 20 ft and the column supports 9 stories with a unit self weight of 150 psf and superimposed dead load of 40 psf, determine the live load that can be assigned to the column if the requirements for factors of safety (ratios of load factors to strength reduction factors) are 2 for the self weight and 3 for the live load.
b. Recommend a spacing of the spiral.

10 Measures of Flexural Response

In Chapters 7 through 9, we studied the response and proportioning of axially loaded members (columns) to develop a perspective for their use in structures. Column response was interpreted in terms of the relationship between axial load and axial deformation. For members subjected to flexure, we shall use a similar approach. The approach we used for axial loading can be summarized in three simple steps.

For several increasing magnitudes of unit strain and for an assumed strain distribution, we determined

1. Unit stress (from $f = E \cdot \varepsilon$ in the linear range of response and from the applicable stress–strain curve—Figure 4.2 or Figure 5.8 in the nonlinear range).
2. The resultants of the internal stresses (in the case of columns resisting concentric axial load: $M = 0$ and $P = \Sigma f \cdot A$).
3. Deflection (from $\delta = L\,\varepsilon$).

In the above list, P is the applied axial load, A is the cross-sectional area of the column, E is Young's modulus, and L is the distance over which change in length is calculated.

Step 1 involves data from coupon tests of like materials and is correct within the limits of relevance of the coupons and their loading conditions. In the nonlinear range of response, the assumed stress–strain curve is used to determine the stress.

Step 2 is based on statics, which has the fewest number of exceptions among all principles used in analyses of load effects.

Step 3 is based on the definition of unit strain.

The development of the load-deflection relationship for a member subjected to bending uses the same concepts involved in these three steps. The reader is expected to have the experience in materials and statics required to apply the concepts involved in the first two steps to the case of elements resisting bending moment. The third step is explained below. A key difference between the procedures followed to understand response to axial load and those for bending moment is related to the determination of deflection from calculated unit strains (step 3), a process that follows the rules of plane geometry. As it was for axial loading, the fundamental assumption is linearity of strain distribution over the cross section of the member. A reinforced concrete section subjected to flexure is likely to develop cracks starting on its tension flange. The presence of the crack raises questions about the meaning of strain in the cracked part of the section.

The photographs in Figure 10.1 show a reinforced concrete beam in four stages of increasing bending moment being applied at its ends. The sense of the applied moments and the visible bending of the beam suggest that the section is subjected to

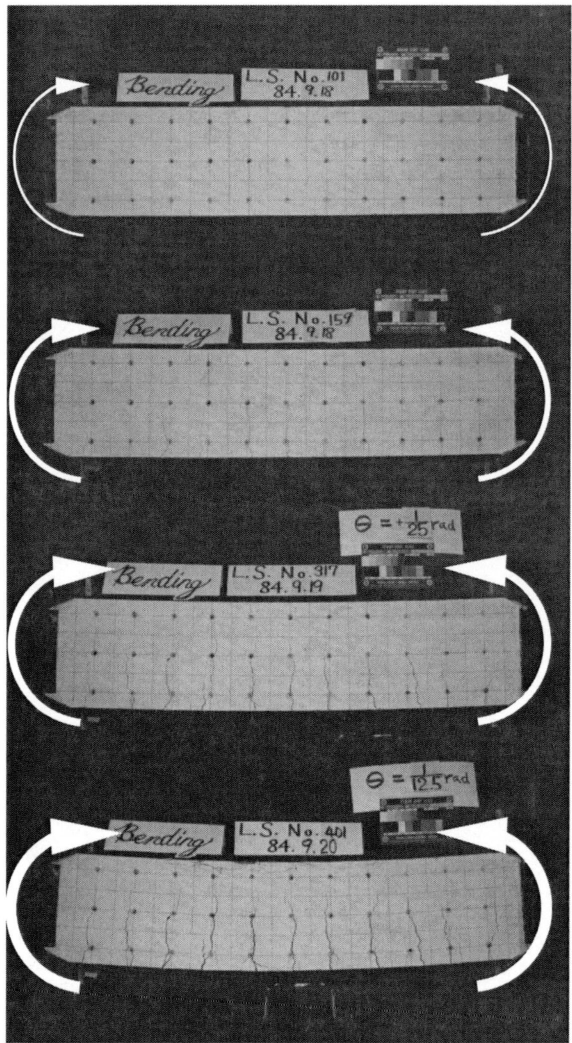

FIGURE 10.1 Beam subjected to bending.

compression at the top and tension at the bottom. The presence of the cracks confirms the inference about the location of the tensile stress but raises the question about continuity of strain. How can one talk of strain at the section of the crack if there is no material to strain? Fortunately, it has been observed that average strain[*] distributions[†] in beams with straight longitudinal axes are very close to linear if the moment remains nearly constant along the gage length and if the tensile reinforcement ratio exceeds 0.2%.[‡] If the reinforcement ratio is below 0.2% in a rectangular section

---

[*] Parallel to the beam axis.

[†] Measured using gage lengths exceeding the crack spacing.

[‡] Typically, beams have tensile reinforcement ratios exceeding 0.2%.

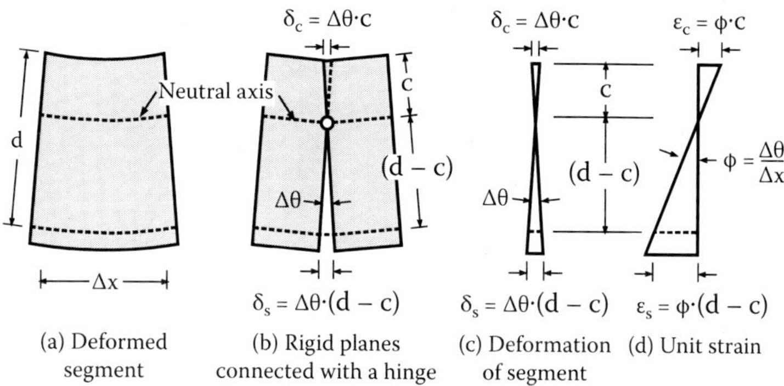

FIGURE 10.2 Definition of unit curvature.

similar to the one shown in Figure 10.1, there may be only one crack in the entire segment, leading to concentrations of deformation in the reinforcement crossing the crack.

As long as the reinforcement ratio is more than 0.2% and flexural cracks form at short intervals, the mean strain distribution over the depth of the beam is close to linear even after cracking of the concrete. The assumption of linear strain distribution over the height of the section, conditionally true, enables the use of elementary mechanics to define/understand the flexural response of reinforced concrete beams.

In dealing with axial load we found the concept of unit strain related to the change in axial shortening very useful in understanding the response of the column. In dealing with bending the parallel concepts are the rate of change in slope (or unit curvature) and deflection. We use those simple concepts by considering the deformation of the small segment in Figure 10.2a. We simplify the deformation as shown in Figure 10.2b, where $\Delta\theta$ is the change in slope. Figure 10.2c shows the distribution of deformations over the height of the section in terms of the change in angle and the distance from the neutral axis. The changes in length, δ_c at the extreme fiber in compression and δ_s at the centroid of the tensile reinforcement, are defined below:

$$\delta_c = \Delta\theta \cdot c \tag{10.1}$$

$$\delta_s = \Delta\theta \cdot (d - c) \tag{10.2}$$

where c is depth to the neutral axis and d is the effective depth (depth from extreme fiber in compression to centroid of tension reinforcement).

The deformations are proportional to the distance from the neutral axis. Dividing the deformations by Δx, we get the unit strains in the extreme fiber in compression and in the tensile reinforcement (Figure 10.2d):

$$\varepsilon_c = \frac{\Delta\theta \cdot c}{\Delta x} = \phi \cdot c \tag{10.3}$$

$$\varepsilon_s = \frac{\Delta\theta \cdot (d - c)}{\Delta x} = \phi \cdot (d - c) \tag{10.4}$$

where unit curvature ϕ is defined as

$$\phi = \frac{\Delta\theta}{\Delta x} \tag{10.5}$$

Note in Figure 10.2d that unit curvature equals the slope of the strain distribution over a section.

$$\varphi = \frac{\varepsilon_c}{c} \tag{10.6}$$

Unit curvature has the unit 1/length and needs to be integrated over a distance in order to have relevance to the physical world. If unit curvature is or may be assumed to be constant over distance Δx (Figure 10.2a), the product

$$\Delta\theta = \phi \cdot \Delta x \tag{10.7}$$

defines the angle change (Figure 10.2b) that occurs over distance Δx.

In the following sections, we shall use the definitions of unit curvature and angle change to understand the flexural response of reinforced concrete beams in linear and nonlinear ranges of response. It is important to note that once we obtain the required strains in the beam, the rest of the process to determine unit curvature and angle change is in the realm of plane geometry. To determine deflection (transverse to the axis of the beam), we shall use other procedures, but they too will be within the principles of plane geometry.

EXAMPLE 10.1

An 18 ft long beam with an 18 × 9 in. cross section is subjected to constant moment (Figure 10.3). The beam is reinforced with a single layer of Grade 60 reinforcing bars with a total area A_s. The bars are located at a depth d = 15 in. from the extreme fiber in compression. Estimate the unit strain and the unit stress in the reinforcement assuming θ = 2 deg (Figure 10.3). Assume the depth to the neutral axis is d/3 = 5 in.

SOLUTION

To estimate strain we first estimate unit curvature. We can estimate unit curvature from the expression $\phi = \Delta\theta/\Delta x$, noting that the total angle change of 2 deg is the result of the unit curvature over half the beam span because the beam slope is zero at mid-span and 2 deg at the support:

$\phi = \Delta\theta/\Delta x = 2 \text{ deg} \times \pi \text{ rad}/180 \text{ deg}/(9 \text{ ft} \times 12 \text{ in./ft}) = 3.5\%/(9 \text{ ft} \times 12 \text{ in./ft})$

$= 0.00032 \text{ in.}^{-1}$

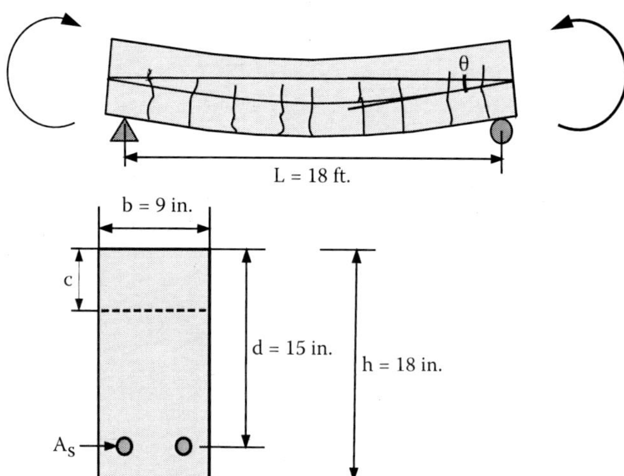

FIGURE 10.3 Example 10.1

The unit strain at the level of the reinforcement is

$$\varepsilon = \phi\,(d - c) = 0.00032\ \text{in.}^{-1} \times (2/3) \times 15\ \text{in.} = 0.0032$$

The calculated unit strain in the reinforcement exceeds the nominal unit strain at yield (0.0021). We conclude that the reinforcement is likely to have yielded and $f_s = f_y = 60$ ksi. Note that our conclusions are independent of the amount of reinforcement in the beam.

BARE ESSENTIALS

- Unit curvature defines the rate at which slope θ changes:

$$\varphi = \frac{\Delta\theta}{\Delta x}$$

- Unit curvature is equal to the rate at which unit strain changes with increases in distance to the neutral axis:

$$\varphi = \frac{\varepsilon_c}{c}$$

- Unit curvature, having a unit of 1/length, needs to be integrated over a finite distance to relate to the physical world.

EXERCISES

1. For the example above, compute the unit strain and unit stress in the reinforcement, assuming c = d/2, c = d/4, and c = d/5. Do you see a trend? If so, describe it in a single sentence.
2. For the example above, compute the maximum compressive unit strain and unit stress in the concrete, assuming c = d/2, c = d/4, and c = d/5. Do you see a trend? If so, describe it in a single sentence. Assume f'_c = 5000 psi.

11 A General Description of Flexural Response

11.1 THE RELATIONSHIP BETWEEN CURVATURE AND BENDING MOMENT

To understand how a beam resists bending moment and how the mechanism of resistance is related to angle change,* we use the metaphor of the crowbar being used to pull a nail from a wall (Figure 11.1). An upward force F applied near the end of the crowbar causes a moment of magnitude Fx at the face of the wall. That moment is balanced by a force couple: T and C. Force C causes compression in the bar. Force T causes tension in the nail and balances force C in the horizontal direction:

$$C = T \tag{11.1}$$

The distance between forces C and T we call jd, where j is a coefficient and d is the distance between the nail and the edge of the crowbar. From the requirement of moment equilibrium:

$$M = F \times x = T \times j \cdot d \tag{11.2}$$

To interpret the relationship between the applied and the resisting moments in the beam shown in Figure 11.2a, consider the equilibrium of the segment to the left of section X. Internal forces must balance the moment caused by the reaction with respect to X (Figure 11.2b). Internal stresses distributed over the section can be represented by concentrated forces C (compression) and T (tension), analogous to the forces acting on the crowbar. Again, equilibrium of forces and moments requires

$$C = T$$

and

$$M = F \times x = T \times j \cdot d$$

The magnitude of the resultants C and T that balance the moment M and the distance between them $(j \cdot d)$ depends on the distribution of unit stresses. The distance $j \cdot d$ is called the internal lever arm.

---

* In Chapter 10 we defined angle change in terms of unit curvature.

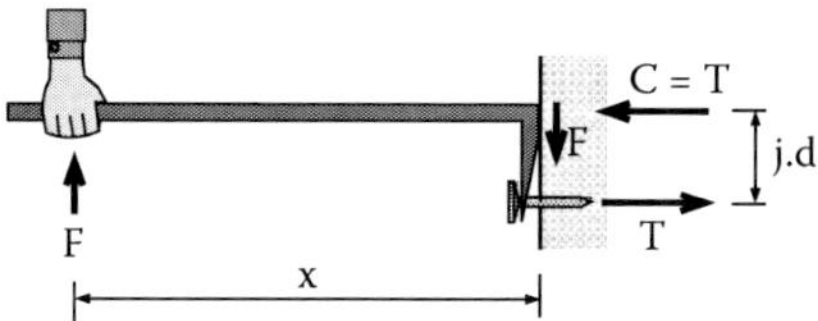

FIGURE 11.1 Crowbar.

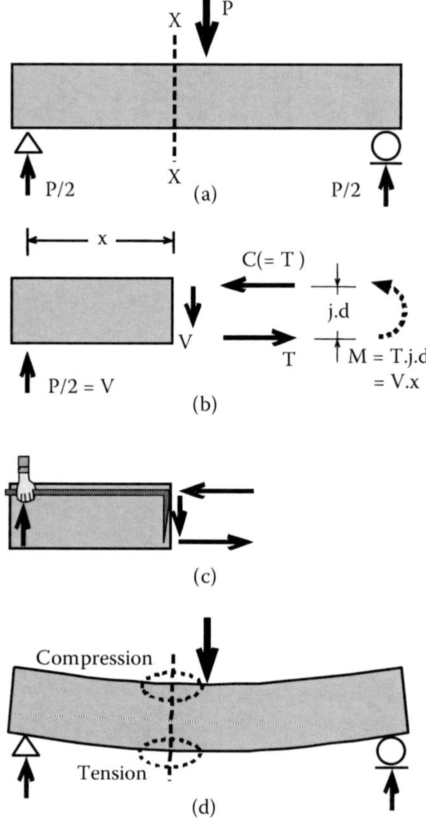

FIGURE 11.2 Internal forces associated with bending.

Equation 11.2 describes how the external moment is related to the resultants of internal normal unit stresses. These stresses are associated with strains. Recalling that, in turn, strain is related to curvature (Chapter 10), we conclude that *there is a relationship between applied moment and angle change.* This relationship can be illustrated using again the crowbar metaphor, as shown in Figure 11.3. In this figure, the mechanisms resisting the normal forces exerted on the wall are represented by two springs. The applied moment causes the bar to rotate. The rotation of the bar causes one spring to stretch and the other to shorten. The deformation of each spring

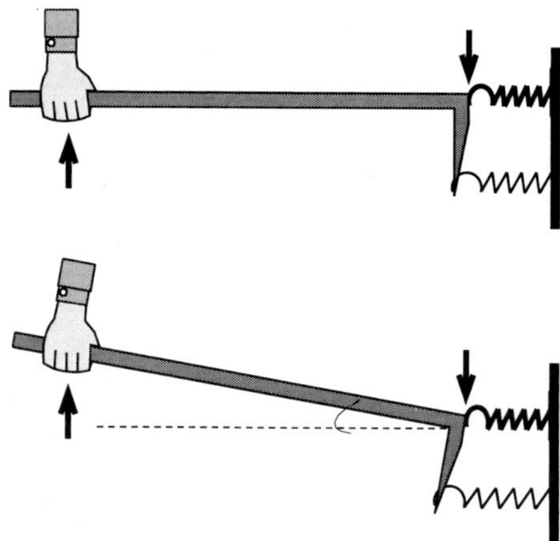

FIGURE 11.3 Crowbar metaphor.

is proportional to forces in the springs. These forces form the couple that balances the applied moment. Moment, bar rotation, and spring deformation are related to one another.

In the beam, angle change is analogous to the amount of rotation of the crowbar. And unit strain is analogous to the unit deformation in the springs. Moment, curvature, and strain are related to one another in the same way that moment, bar rotation, and spring deformation are related.

11.2 STAGES OF RESPONSE

To describe the relationship between moment, unit curvature, and unit strain quantitatively, we consider a beam cross section with a single layer of tensile reinforcement (Figure 11.4). If this beam is subjected to bending as shown in Figure 11.5, and if the ratio of area of reinforcement to the product bd exceeds approximately 0.2% (Chapter 16), the beam goes through four different stages of response, depending on the amount of reinforcement.

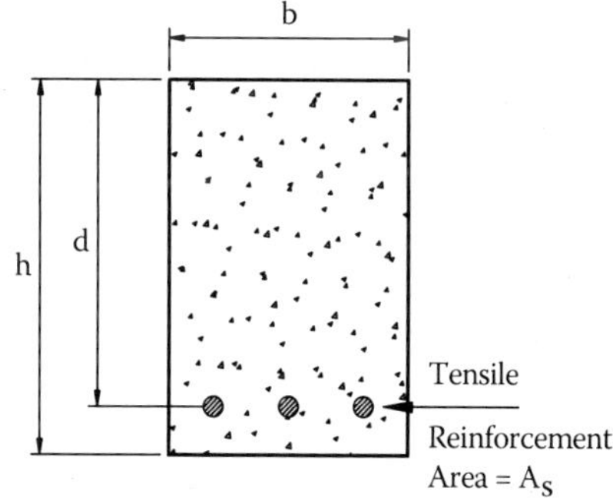

FIGURE 11.4 Cross section of beam with single layer of reinforcement.

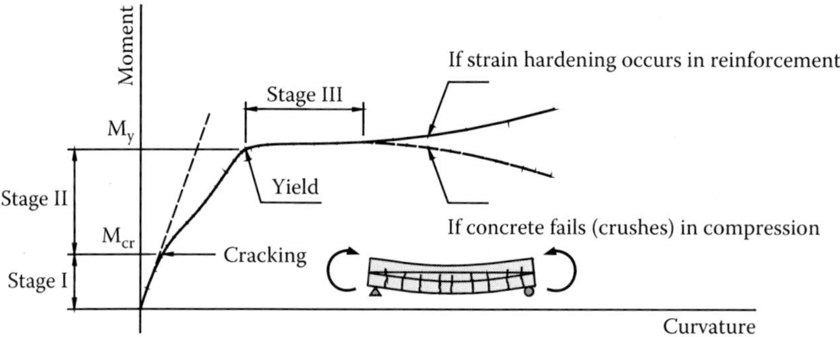

FIGURE 11.5 Typical moment–deflection relationship for beam subjected to bending.

Initially, the beam responds linearly (an increase in moment is directly proportional to an increase in displacement) until the tensile strength of the concrete is reached and it cracks at moment M_{cr} (Stage I in Figure 11.5). As the moment is increased beyond M_{cr} (Stage II in Figure 11.5), stiffness decreases; i.e., the rate of change of moment with curvature decreases.

When the reinforcement reaches its yield stress, there is a strong change in stiffness (Stage III, Figure 11.5). Deformation increases with virtually no change in moment. Depending on the amount of tensile reinforcement, the moment may start increasing again (Figure 11.5), or it may start decreasing as indicated by the broken curve in Figure 11.5.

The goal in flexural design is, given a moment demand, to determine the proper size of the section and the amount of reinforcement. In Chapters 13 and 14 we are going to do that in accordance with mechanics. To accomplish that goal, we need to commit to assumptions about the strain distribution over the depth of the section and the stress-strain properties of the particular type of concrete and reinforcement we intend to use. In this section, we shall short-circuit all that and see if we can accomplish the same within the realm of good engineering judgment.

The depth, h, of the section is usually determined as a fraction of the span, and for a continuous beam, it is often taken as the span length in feet converted to inches. For example, the starting value of the beam depth for a span of 24 ft would be 24 in. Please note that we call it a starting value. For specific cases, determination of dimensions may require several repetitions to obtain the optimum.

The width of the section for a rectangular beam is usually 1/2 to 2/3 of its depth, but depending on the special conditions of the beam in question, one may deviate from it by as much as 50%, provided the requirements of the applicable code are satisfied.

In determining the amount of reinforcement, the engineer needs to be strongly conscious of the reinforcement ratio, the ratio of the total cross-sectional area of reinforcement in tension to the product of the beam width and effective depth, bd (see Figure 11.6). For a beam, this ratio ought not to be below 0.2% and ought not to exceed 2.5%. These bounds are driven not only by strength and behavioral considerations, but also by construction demands.

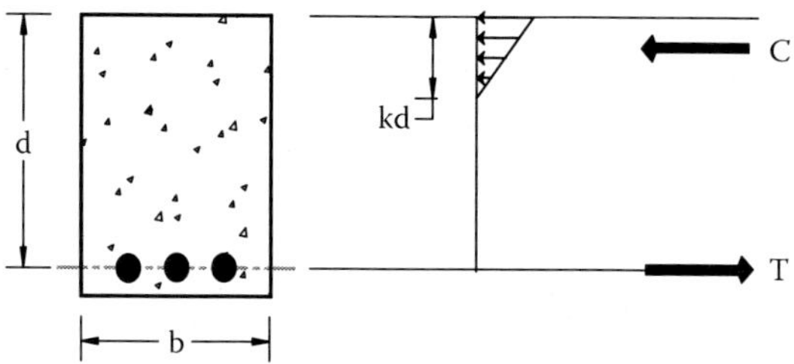

FIGURE 11.6 Resultants of internal stresses.

It is clear from our discussion of the relationship between external moment and internal resistance that if we can estimate the dimension jd, we can determine the internal forces and that would allow us to estimate the amount of reinforcement required as well as decide whether the concrete quality selected for design will be adequate. So the flexural challenge reduces to a good if not exact estimate of the coefficient j. In the following text we make an attempt to do that, assuming concrete stress remains in the linear range. The reader is encouraged to see whether a better result can be achieved by a simpler and faster procedure and by using different assumptions.

If concrete remains in its linear range of response, with its tensile strength ignored, internal stresses should be as pictured in Figure 11.6. The neutral axis should be somewhere near the mid-depth of the section. If concrete stress varies linearly from zero at the neutral axis to a maximum at the extreme fiber in compression, the centroid of the compressive force would be at a distance of kd/3 from the extreme fiber in compression. The stress in the reinforcement should be limited by the yield stress. And we can tell that the more the reinforcement, the higher would be the distance kd in order for the compressive force to balance the tensile force. But to know T or C, we need to know j. To know j, we need to know kd, and to know kd, we need to know T or C. It is a vicious cycle of sorts. An experienced engineer would make a good guess for j and solve for C and T by dividing M by jd. But we do not have the experience. How can we do it?

We take a brief look at how coefficient j would vary with coefficient k with the prejudgment that $k > 0.6$ would be unlikely. For the linear stress distribution we have assumed

$$j = 1 - \frac{k}{3}$$

We infer from the plot in Figure 11.7 that 0.9 would be a good value to take for j. For the assumed linear distribution of stress, the maximum error would not exceed 11% and the expected error, believing that k would be somewhere in the range 0.1 to 0.5, would not exceed 5%. The reader is encouraged to investigate how our estimate of j would vary for different stress distributions.

Given a moment demand M, now we are ready to estimate absolute values of C and T,

$$C = T = \frac{M}{0.9d}$$

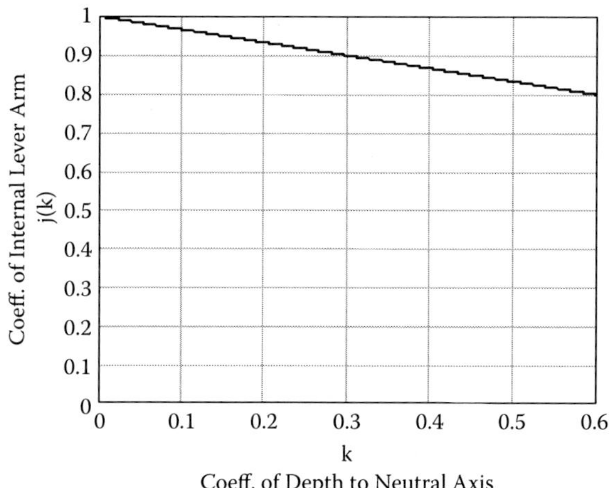

FIGURE 11.7 Relationship between coefficients j and k.

EXAMPLE 11.1

For the reinforced concrete beam shown in Figure 11.8, and for the following parameters, compute the maximum unit stress in the reinforcement.

$$L = 24 \text{ ft}$$

$$P = 40 \text{ kip}$$

$$d = 24 \text{ in.}$$

We assume that (1) *all* internal tensile forces are resisted by the reinforcement, (2) the resultant of compressive forces acts at a distance of 0.1d from the top face of the beam, (3) the total cross-sectional area of reinforcement is $A_s = 3$ in.^{2}, and (4) self weight is negligible.

SOLUTION

The maximum moment takes place at mid-span. Its magnitude is

$$M = PL/4 = 40 \text{ kip} \times 24 \text{ ft}/4 = 240 \text{ kip-ft}$$

Internal forces T and C balance the moment M:

$$M = T \times jd = C \times jd$$

The location of the resultant of compressive stresses given implies jd = d − 0.1d = 0.9d. Using j = 0.9 and solving the previous expression for T,

$$T = M/(jd) = 240 \text{ kip-ft} \times 12 \text{ in./ft}/(0.9 \times 24 \text{ in.}) = 133 \text{ kip}$$

The unit stress is

$$f_s = T/A_s = 133 \text{ kip}/3 \text{ in.}^2 = \boxed{44 \text{ ksi}}$$

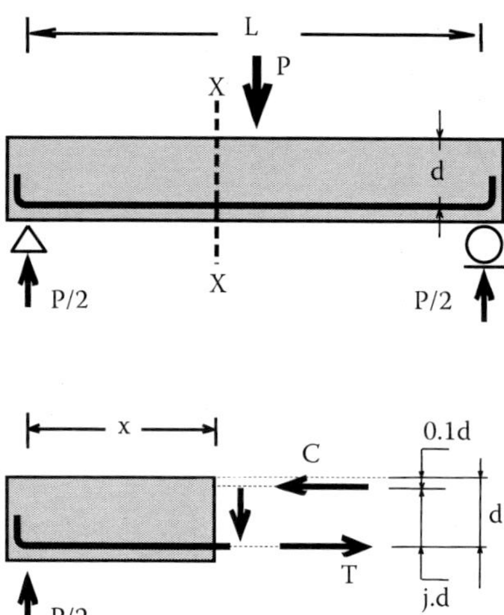

FIGURE 11.8 Example.

> **BARE ESSENTIALS**
>
> Bending moment is resisted by an internal couple formed by the resultants of compression and tension stresses C and T.
>
> The distance between these resultants is called the internal lever arm and is expressed as a product of a coefficient j and the distance d from resultant T to the extreme fiber in compression.
>
> Coefficient j can be assumed to be 0.9 to determine initial proportions.

EXERCISES

1. For the conditions described in the previous example, compute the maximum load that can be resisted by the beam. Assume the reinforcing bars used are Grade 60, and assume that the strength of the concrete in compression does not limit the flexural strength of the section.
2. Reproduce the plot in Figure 11.7 assuming that the distribution of compressive stresses in the concrete is parabolic. Ignore again tensile stresses in the concrete.

12 Moment–Curvature Relationship before Flexural Cracking

In this section, we focus on the response of an uncracked rectangular reinforced concrete section to applied moment (Stage I as defined in Chapter 11). A reinforced concrete beam in a structure may have cracks caused by shrinkage and temperature effects, even if it has not been subjected to moments high enough to cause flexural cracking. Furthermore, self weight may suffice to crack some of the beam sections. This does not mean we should ignore the behavior of the uncracked section because there are uncracked regions in a cracked beam and the overall response of the beam in flexure depends on the responses of the uncracked as well as the cracked sections.

To develop a simple relationship between moment and curvature for a rectangular section (Figure 12.1a) of a straight beam, we make two pivotal assumptions:

1. The distribution of unit strain over the section is linear (Figure 12.1b).
2. The unit stress changes linearly with unit strain (Figure 12.1c).

It is relevant to note that both assumptions have their roots in observation. Short of explaining it in terms of molecular movement, strain linearity is justified by observation. The measure of stiffness of concrete (Young's modulus, E_c) is also based on observation.

Considering the natural scatter expected in the values of Young's modulus (as well as assuming it to be the same in tension and in compression), it is counterproductive to consider the effect of the reinforcement on flexural response.*

Because we have assumed the same modulus in tension and compression and ignored the reinforcement, we recognize that the neutral axis (position with zero strain) will be at mid-height of the rectangular section. If the stress in the extreme fiber in compression is f_c, the stress at a distance y from the neutral axis is

$$f(y) = \frac{y}{\frac{1}{2}h} f_c \tag{12.1}$$

* Even though the presence of tensile reinforcement may appear to increase the cracking moment, it decreases it because shrinkage of the concrete is restrained by the reinforcement. As long as a beam is reinforced moderately with a reinforcement ratio not exceeding 1.5%, the reinforcement effect on stiffness can be neglected.

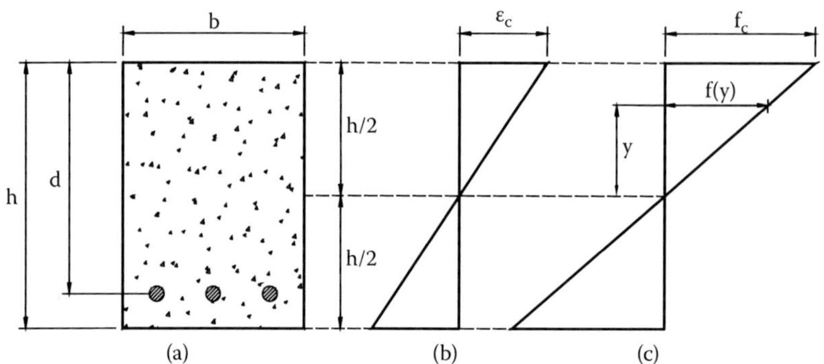

FIGURE 12.1 (a) Cross section. (b) Unit strain. (c) Unit stress.

The unit stress f_c acts on a differential of area (Figure 12.1 and Figure 12.2),

$$dA = b \cdot dy \tag{12.2}$$

and is associated with a differential of force,

$$dF = f(y) \cdot dA \tag{12.3}$$

The resultant of compressive stresses is

$$C = \int_0^{\frac{1}{2}h} f(y) \cdot dA = \frac{1}{2} f_c b \frac{h}{2} \tag{12.4}$$

The product $f(y_n)dA$ can also be obtained by treating it as a differential of volume (Figure 12.2). In that case the magnitude of the force C is the volume of the upper wedge in Figure 12.2.

Because the beam is *not* subjected to axial load, C must equal the resultant of tensile stresses T:

$$C = T = \frac{1}{4} b \cdot h \cdot f_c \tag{12.5}$$

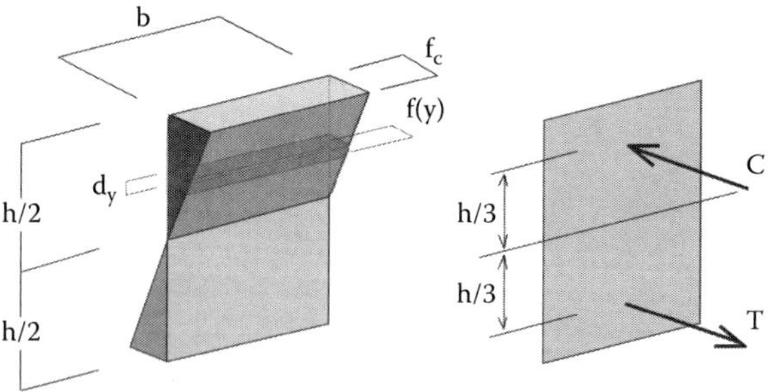

FIGURE 12.2 Unit stress distribution before cracking.

The lines of action of these resultants pass through the centroids of the volumes shown in Figure 12.2. The distance between the two centroids is

$$j \cdot d = \frac{2}{3} \cdot h \tag{12.6}$$

Therefore,

$$M = T \cdot \frac{2}{3} h = \frac{1}{6} b \cdot h^2 \cdot f_c = S \cdot f_c \tag{12.7}$$

Equation 12.7 relates the maximum stress in the section to the applied moment. Its form is a result of the assumed distribution of unit stress. Unit stress distributions with different shapes will lead to different results. The quantity $bh^2/6$ is called the section modulus S (for a rectangular section) and is equal to the ratio of moment of inertia of the full section divided by the distance from the neutral axis to the point where stress is maximum (in this case, $c = h/2$). The section modulus is a function of the shape of the cross section.

We compute the maximum compressive strain as

$$\varepsilon_c = \frac{f_c}{E_c} = \frac{M}{E_c S} \tag{12.8}$$

and unit curvature as

$$\varphi = \frac{\varepsilon_c}{c} = \frac{M}{E_c I_g} \tag{12.9}$$

where $I_g = \frac{1}{12} bh^3$ and $c = h/2$.

The boundary between Stages I and II is given by the moment at cracking M_{cr}. For normal weight concrete, a reasonable lower bound to the moment at cracking can be expressed as

$$M_{cr} = S \cdot f_r = \frac{1}{6} b \cdot h^2 \cdot f_r = b \cdot h^2 \cdot \sqrt{f_c'} \tag{12.10}$$

Here f_r is modulus of rupture and is assumed to be $f_r = 6\sqrt{f_c'}$, although it can reach larger values (Chapter 5).

EXAMPLE 12.1

For the reinforced concrete beam shown in Figure 12.3, and for the following parameters, compute the magnitude of the uniformly distributed load w expected to cause cracking. Ignore self weight.

$$f_c' = 5000 \text{ psi}$$

$$L = 18 \text{ ft}$$

$$h = 18 \text{ in.}$$

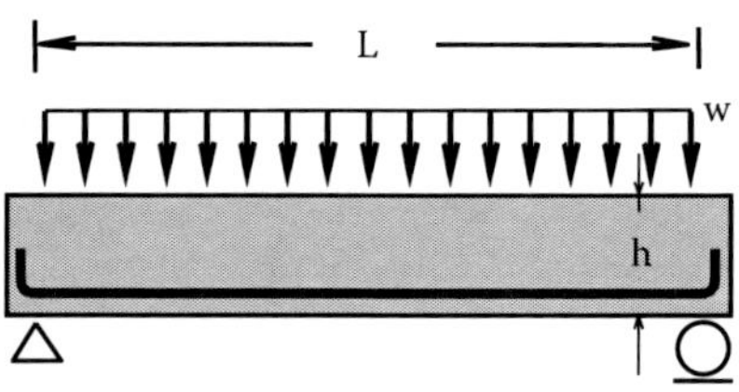

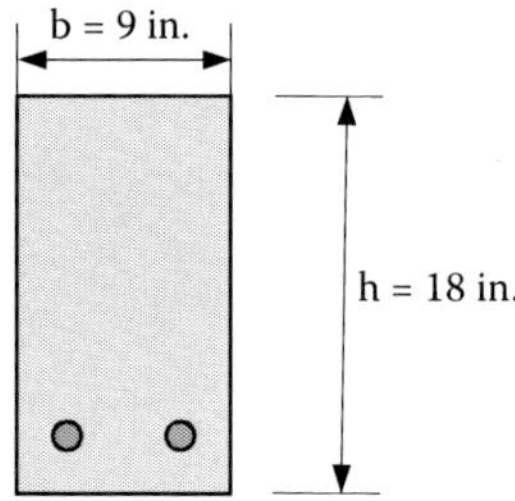

FIGURE 12.3 Example.

SOLUTION

The maximum "applied" moment is located at mid-span and is equal to

$$M_a = wL^2/8$$

The moment of inertia of the full section, corrected to three significant figures, is

$$I_g = 9 \text{ in.} (18 \text{ in.})^3/12 = 4370 \text{ in.}^4$$

The section modulus is

$$S = I_g/(h/2) = 4370 \text{ in.}^4/9 \text{ in.} = 486 \text{ in.}^3$$

Taking the modulus of rupture as

$$f_r = 6\sqrt{(5000)} \text{ psi} = 420 \text{ psi}$$

we obtain

$$M_{cr} = S \cdot f_r = 486 \text{ in.}^3 \times 420 \text{ psi} \times (1 \text{ kip}/1000 \text{ lbf}) \times (1 \text{ ft}/12 \text{ in.}) = 17 \text{ kip-ft}$$

Cracking is expected to occur when the applied moment reaches the moment at cracking ($M_a = M_{cr}$). Solving for w,

$$w = 8\ M_{cr}/L^2 = 8 \times 17 \text{ kip-ft}/(18 \text{ ft})^2 = 0.42 \text{ kip-ft}$$

BARE ESSENTIALS

Before cracking—in Stage I—the relationship between curvature and bending moment is

$$\phi = \frac{\varepsilon_c}{c} = \frac{M}{E_c I_g}$$

Response remains in Stage I as long as the applied moment does not exceed

$$M_{cr} = S \cdot f_r$$

For beams with rectangular cross sections, a reasonable lower bound to the moment at cracking can be expressed as

$$M_{cr} = S \cdot f_r = b \cdot h^2 \cdot \sqrt{f'_c}$$

EXERCISE

For the reinforced concrete beam shown in Figure 12.4, and for the following parameters, estimate the load expected to cause cracking. Ignore self weight. (Keep in mind that S in the expression $M_{cr} = S \cdot f_r$ is a function of the shape of the cross section and is equal to the ratio of moment of inertia of the full section divided by distance from the neutral axis to the point where stress is maximum.)

$$f'_c = 5000 \text{ psi}$$

$$L = 18 \text{ ft}$$

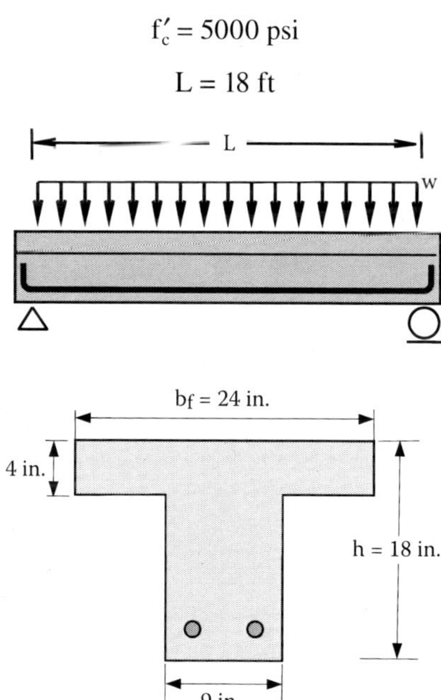

FIGURE 12.4 Exercise.

13 Linear Response of Cracked Sections

In this chapter, we continue focusing on a rectangular section reinforced in tension. Cracking causes a rapid reduction in the moment resisted by the section. This reduction is compensated by the action of the reinforcement as the response changes from Stage I to Stage II. In Stage II, we ignore tensile unit stresses in the concrete, but we retain the assumptions we made in Stage I for the strain distribution and the relationship between compressive unit strain and unit stress in the concrete. The depth to the neutral axis of the section with a flexural crack is expressed as kd (Figures 11.6 and 13.1), where d is the depth to the centroid of the tensile reinforcement from the extreme fiber in compression (the effective depth) and k is a coefficient that depends on the amount of reinforcement and the relative stiffness of steel to that of concrete.

From the geometry of the assumed strain distribution (Figure 13.1), we obtain the relationship

$$\frac{\varepsilon_s}{\varepsilon_c} = \frac{1-k}{k} \tag{13.1}$$

where ε_s is the unit strain in reinforcement, ε_c is the unit strain in concrete at the extreme fiber in compression, and k is the ratio of neutral axis depth from extreme fiber in compression to the effective depth of the reinforcement d.

From equilibrium of normal forces acting on the cross section (Figure 13.1),

$$T = A_s \cdot f_s = \frac{1}{2} \cdot f_c \cdot b \cdot k \cdot d = C \tag{13.2}$$

where A_s is the total cross-sectional area of tensile reinforcement and f_s is the unit stress in tensile reinforcement.

To simplify and generalize the expressions and to provide a useful index value to understand the effectiveness of tensile reinforcement in a section, we define *reinforcement ratio*, ρ, as the ratio of the total area of the tensile reinforcement, A_s, to the product of b, the width of the zone in compression, and d, effective depth:

$$\rho = \frac{A_s}{b \cdot d} \tag{13.3}$$

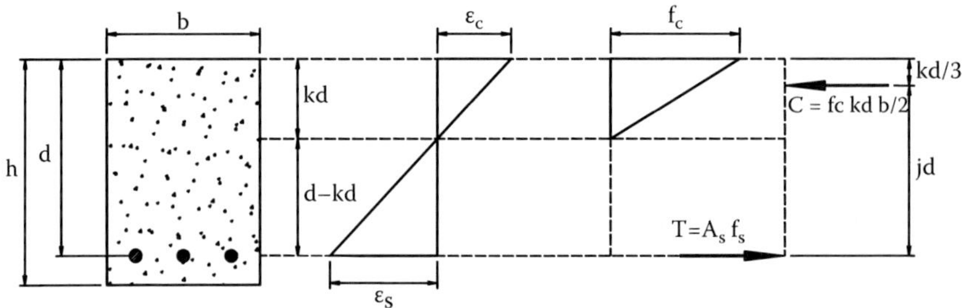

FIGURE 13.1 Strain and stress distributions in Stage II. (a) Section. (b) Unit strain. (c) Concrete. (d) Resultant unit stress.

In beams, the reinforcement ratio ρ is usually in between 0.5 and 2.5%. In slabs, it can be as low as 0.2%. *Ratios x in the range 1% to 1.5% are common.* Replacing A_s in Equation 13.2 using the definition in Equation 13.3:

$$\rho \cdot f_s = \frac{1}{2} \cdot f_c \cdot k \tag{13.4}$$

Using the material moduli E_s and E_c, we rewrite Equation 13.4 in terms of unit strain and the modular ratio $n = E_s/E_c$:

$$\rho \cdot \varepsilon_s \cdot n = \frac{1}{2} \cdot \varepsilon_c \cdot k \tag{13.5}$$

or

$$\frac{\varepsilon_s}{\varepsilon_c} = \frac{k}{2 \cdot \rho \cdot n} \tag{13.6}$$

Equating the right-hand terms of Equations 13.1 and 13.6,

$$\frac{1-k}{k} = \frac{k}{2 \cdot \rho \cdot n} \tag{13.7}$$

We solve for k and take the positive root to obtain the coefficient k for a rectangular section reinforced in tension only:

$$k = \sqrt{\rho^2 \cdot n^2 + 2 \cdot \rho \cdot n} - \rho \cdot n \tag{13.8}$$

The value of k is *usually* between 0.2 and 0.4. Equation 13.8 gives us a convenient vehicle to calculate the neutral axis depth in Stage II of a cracked rectangular section with tension reinforcement.

Taking moments about the centroid of the compressive force,

$$M = A_s \cdot f_s \cdot d \cdot \left(1 - \frac{k}{3}\right) \tag{13.9}$$

This expression allows us to estimate the reinforcement unit stress in Stage II with the caveat that the reinforcement unit stress is likely to be less than that calculated because we have neglected the contribution from concrete in tension. Our results indicate that, in Stage II, the internal lever arm jd is equal to

$$j \cdot d = \left(1 - \frac{k}{3}\right) \cdot d \tag{13.10}$$

and

$$f_s = \frac{T}{A_s} = \frac{M}{j \cdot d \cdot A_s} \tag{13.11}$$

with f_s varying in direct proportion to the applied moment.

Unit curvature can be expressed as

$$\phi = \frac{\varepsilon_s}{(1-k) \cdot d} \tag{13.12}$$

where $\varepsilon_s = f_s/E_s \leq \varepsilon_y$.

We must remember that we have assumed the concrete unit stress to be linearly related to concrete strain. This assumption needs to be checked. We can do so by using the definition of unit curvature:

$$\phi = \frac{\varepsilon_c}{k \cdot d} \tag{13.13}$$

From this expression, and assuming the concrete to be linear, we obtain

$$f_c = E_c \cdot \phi \cdot k \cdot d \tag{13.14}$$

If f_c does not exceed $f'_c/2$, the error associated with the nonlinearity of concrete may be considered to be negligible.

Unit curvature can also be estimated as

$$\phi = \frac{M}{E_c \cdot I_{cr}} \tag{13.15}$$

$$I_{cr} = I_{concrete} + I_{steel} = \left[\tfrac{1}{3} b \cdot (kd)^3\right] + n \cdot A_s \cdot (d - kd)^2 \tag{13.16}$$

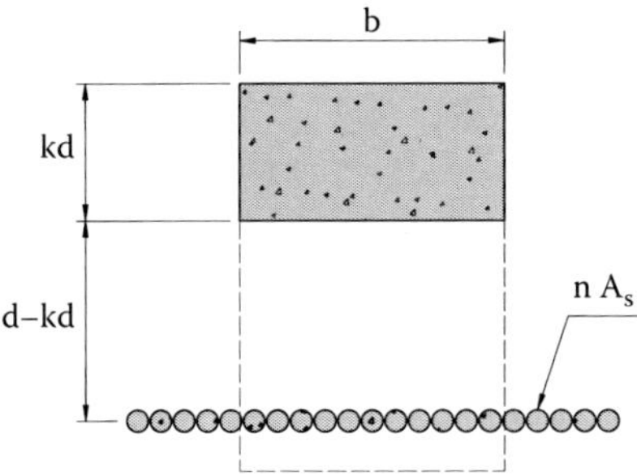

FIGURE 13.2 Transformed section.

This expression represents the moment of inertia of the transformed section shown in Figure 13.2. The first term on the right-hand side of this expression is the moment of inertia, about the neutral axis, of the rectangular area of concrete above it. The second term is the moment of inertia, about the neutral axis, of an area equal to n times A_s. This moment of inertia can be obtained using the parallel axis theorem, neglecting the moment of inertia of the area nA_s with respect to its own centroidal axis, and recognizing that the neutral axis is at the centroid of the transformed section. Equation 13.15 is also based on the assumption that the materials are linear. Its applicability should be checked by comparing the peak stresses in the section to the limits mentioned previously.

EXAMPLE 13.1

For the section shown in Figure 13.3, compute the stress in the reinforcement associated with a moment of 60 kip-ft. Assume A_s = 1.58 in² (two #8 Grade 60 bars), f'_c = 5000 psi, and E_c = 4000 ksi.

SOLUTION

We check first whether the section is in Stage I or II:

$$M_{cr} = 1/6\ (9\ \text{in.})\ (18\ \text{in.})^2\ (6 \times \sqrt{5000}\ \text{psi}) \times (1\ \text{kip}/1000\ \text{lbf}) \times (1\ \text{ft}/12\ \text{in.})$$

$$= 17\ \text{kip-ft} < 60\ \text{kip-ft}$$

We conclude that the section is cracked at M = 60 kip-ft. For ρ = 1.58 in.²/(9 in. × 15 in.) = 1.2% and n = 29,000 ksi/4000 ksi = 7.3, the relative depth to the neutral axis is

$$k = \sqrt{0.012^2 \cdot 7.3^2 + 2 \cdot (0.012) \cdot (7.3)} - (0.012) \cdot (7.3) = 0.34$$

and the internal lever arm is j = 1 – 0.34/3 = 0.9.

The unit stress in the reinforcement is therefore

$$f_s = \frac{60\ \text{kip} \cdot \text{ft} \cdot 12 \dfrac{\text{in.}}{\text{ft}}}{1.58\ \text{in.}^2 \cdot 0.9 \cdot 15\ \text{in.}} = 34\ \text{ksi}$$

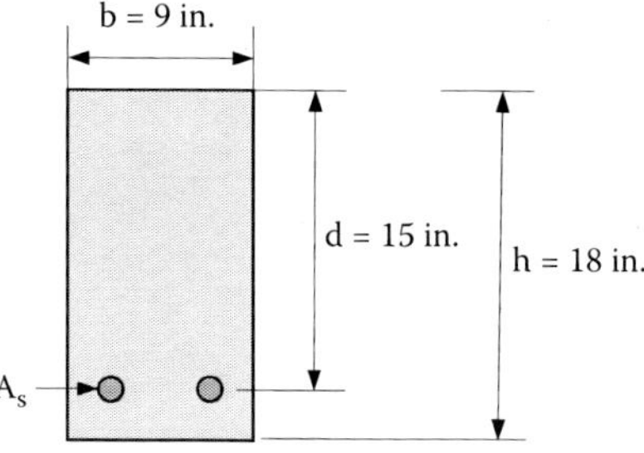

FIGURE 13.3 Example 13.1.

The computed stress is less than the yield stress of the reinforcing bars, confirming that the section is in its linear range of response. The reader is asked to check whether the stress in the concrete is within the limits of applicability of the formulation used.

EXAMPLE 13.2

For the section shown in Figure 13.4, compute the curvature and the stress in the reinforcement associated with a moment of 60 kip-ft. Assume A_s = 1.58 in^2 (two #8 bars), f_c' = 5000 psi, and E_c = 4000 ksi. Assume the section to be cracked.

SOLUTION 1

We first assume that the concrete in compression is within the flange. For ρ = 1.58 in.^{2}/(18 in. × 15 in.) = 0.6% and n = 29,000 ksi/4000 ksi = 7.3, the relative depth to neutral axis is

$$k = \sqrt{0.006^2 \cdot 7.3^2 + 2 \cdot (0.006) \cdot (7.3)} - 0.006 \cdot 7.3 = 0.26$$

The computed depth to the neutral axis is 0.26 × 15 in. = 3.9 in. < 4 in., which supports our assumption about the location of the neutral axis and allows us to treat the section as if it were rectangular.

The stress in the reinforcement is

$$f_s = \frac{M}{A_s \cdot (1 - k/3) \cdot d} = \frac{60 \text{ kip} \cdot \text{ft} \cdot 12 \frac{\text{in.}}{\text{ft}}}{A_s \cdot (1 - 0.26/3) \cdot 15 \text{ in.}} = 33 \text{ ksi}$$

Again, the stress is smaller than the yield stress. We observe that the change in the stress in the reinforcement caused by the effect of the flange is small. The reader is asked to check whether the stress in the concrete is within the limits of applicability of the stress state used.

The strain associated with this stress is

$$\varepsilon = \frac{f_s}{E_s} = \frac{33 \text{ ksi}}{29,000 \text{ ksi}} = 0.0011$$

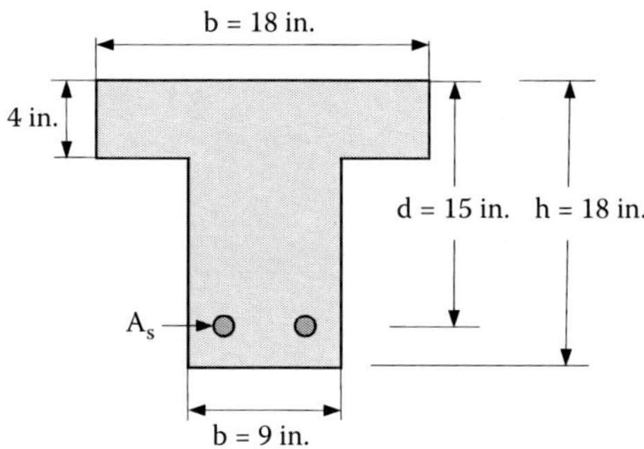

FIGURE 13.4 Example 13.2.

and the associated unit curvature is

$$\phi = \frac{\varepsilon_s}{(1-k)\cdot d} = \frac{0.0011}{(1-0.26)\cdot 15\text{ in.}} = 0.0001\frac{1}{\text{in.}}$$

SOLUTION 2

As shown in Solution 1, the depth to the neutral axis is 0.26 × 15 in. = 3.9 in. < 4 in. (and we therefore treat the section as if it were rectangular).

The moment of inertia of the cracked section is

$$I_{cr} = I_{concrete} + I_{steel} = \left[\tfrac{1}{3}18\text{ in.}\cdot(3.9\text{ in.})^3\right]$$

$$+ 7.3\cdot 1.58\text{ in.}^2\cdot(15\text{ in.} - 3.9\text{ in.})^2$$

$$\approx 355\text{ in.}^4 + 1420\text{ in.}^4 = 1775\text{ in.}^4$$

The curvature is

$$\phi = \frac{M}{E_c I_{cr}} = \frac{60\text{ kip}\cdot\text{ft}\times 12\frac{\text{in.}}{\text{ft}}}{4000\text{ ksi}\cdot 1775\text{ in.}^4} = 0.0001\frac{1}{\text{in.}}$$

Using this estimate for curvature, we can compute the strain in the reinforcement as

$$\varepsilon_s = \phi\cdot(15\text{ in.} - 3.9\text{ in.}) = 0.0001\frac{1}{\text{in.}}\cdot(15\text{ in.} - 3.9\text{ in.}) = 0.0011$$

and stress as

$$f_s = E_s\cdot\varepsilon_s = 29000\text{ ksi}\cdot 0.0011 = 32\text{ ksi}$$

This result is, for all practical purposes, the same as the one above. The difference is related to rounding. Nevertheless, solution 1 may be simpler to understand, as it does not require abstract transformations of the cross section into fictitious equivalents.

BARE ESSENTIALS

- After cracking—in Stage II—the relationship between curvature and bending moment is

$$\phi = \frac{\varepsilon_c}{k \cdot d} = \frac{\varepsilon_s}{(1-k) \cdot d} = \frac{M}{E_c \cdot I_{cr}}$$

and reinforcement unit stress is linearly proportional to applied moment:

$$M = A_s \cdot f_s \cdot jd$$

with $j = (1 - k/3)$ and

$$k = \sqrt{\rho^2 \cdot n^2 + 2 \cdot \rho \cdot n} - \rho \cdot n$$

- Common values: r = 1% to 1.5%, k = 0.3 to 0.4, j = 0.9, $I_{cr} \cong 0.3\, I_g$.
- For a rectangular section with tensile reinforcement ratio of 1 to 1.5%, unit curvature at yield (f_y = 60 ksi) is approximately 0.002/(0.6d).

EXERCISES

1. Compute the maximum stress in the concrete for Examples 13.1 and 13.2. What is your main conclusion, expressed in one sentence?
2. At the applied moment of 60 kip-ft, can the concrete be assumed to be linear in Examples 13.1 and 13.2?
3. Estimate the cracking moment for the section defined in Example 13.2.
4. For the section defined in Example 13.2, compute the unit curvature and the unit stress in the reinforcement associated with a moment of 40 kip-ft. Compare your answer with that obtained in the example for 60 kip-ft. What do you infer from this comparison?

14 Limiting Moment and Unit Curvature

The moment–curvature relationship of a reinforced concrete section having a reinforcement ratio not exceeding 2% can be defined adequately by three pairs of coordinates (Figure 14.1). These break points refer to (1) cracking, (2) yielding, and (3) useful limit. How the coordinates for points (1) and (2) can be determined was discussed in the preceding two chapters. For the assumed shapes of the curves of the concrete and the steel (Figure 14.2), the stress conditions at the three defining points are illustrated qualitatively in Figures 14.2 and 14.3. Point (3) refers to the useful limit of curvature. Unless the tensile reinforcement ratio is extremely low or the reinforcement material is brittle, the limit to the moment–curvature relationship is governed by the useful limit of compressive strain in the concrete. The reason for invoking the qualifier "useful" is that the measurements of the limiting compressive strain in concrete have varied from 0.003 to 0.010, depending on how and where it is measured. The strain limit assumed depends on the purpose of the calculation made and the judgment of the engineer related to acceptable risk. The limit for concrete compressive strain is a chosen limit and not an absolute measure, as it may be for concrete strength.

14.1 A SIMPLE PROCEDURE FOR DETERMINING THE LIMIT TO THE MOMENT–CURVATURE RELATIONSHIP

Before we develop a detailed solution for M_u and ϕ_u, it is instructive to present an approximate solution for a beam of a rectangular cross section. To obtain an approximate solution for the useful limit, we assume

1. The tensile reinforcement is at the yield stress.
2. The mean stress in the compressed concrete above the neutral axis is $0.7f'_c$.
3. The centroid of the total normal force in the concrete is at 0.4c from the extreme fiber in compression (c being the depth to the neutral axis as indicated in Figure 14.4).

From equilibrium of forces acting on the section, we determine the depth of the neutral axis.

$$\text{Tensile force} = \text{Compressive force}$$

$$A_s f_y \approx 0.7 f'_c b k_u d \tag{14.1}$$

$$k_u \approx \frac{\rho \cdot f_y}{0.7 \cdot f'_c} \tag{14.2}$$

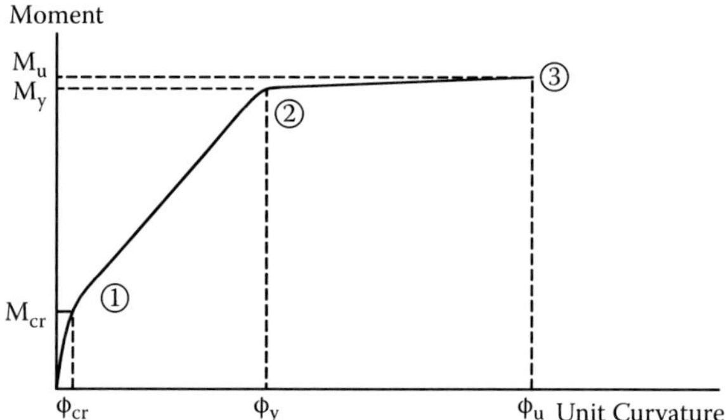

FIGURE 14.1 Moment–curvature relationship.

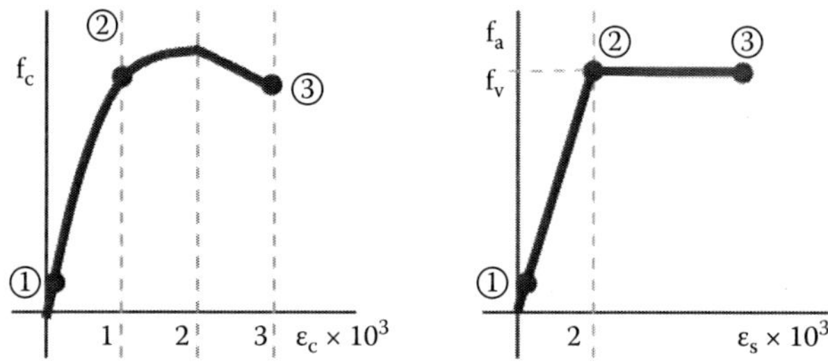

FIGURE 14.2 Stress–strain curves of concrete and steel.

where A_s is the cross-sectional area of tensile reinforcement, f_y is the yield stress of tensile reinforcement, f_c' is the compressive strength (cylinder) of concrete, b is the width of the rectangular section, k_u is the ratio of the depth of the neutral axis c to the effective depth of the section, d is the effective depth of the beam, and ρ is the reinforcement ratio (A_s/bd).

Although it is seldom necessary for moderate amounts of reinforcement in beams (0.2% ≤ ρ ≤ 2%), we check to make certain the yield strain has been reached. Assuming that the useful limit of compressive strain controls and that strain is distributed linearly,

$$\varepsilon_{su} = \varepsilon_{cu} \cdot \frac{1 - k_u}{k_u} \tag{14.3}$$

where ε_{su} is the unit strain in reinforcement at the limiting condition and ε_{cu} is the useful limit of the compressive strain of concrete (may be assumed to be 0.003 to 0.004, depending on the desired degree of conservatism). This limit is set to reflect the fact that the capacity of a beam to bend and resist the bending moment may be limited by the capacity of the concrete to deform in compression (Figure 14.5). k_u is the ratio of the depth to the neutral axis to the effective depth.

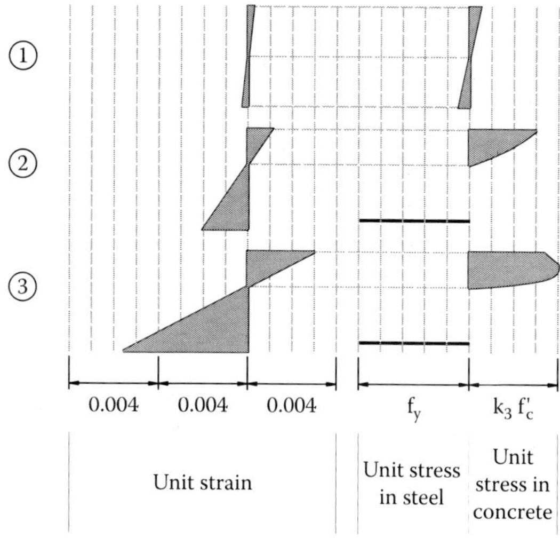

FIGURE 14.3 Strain and stress conditions at three defining points.

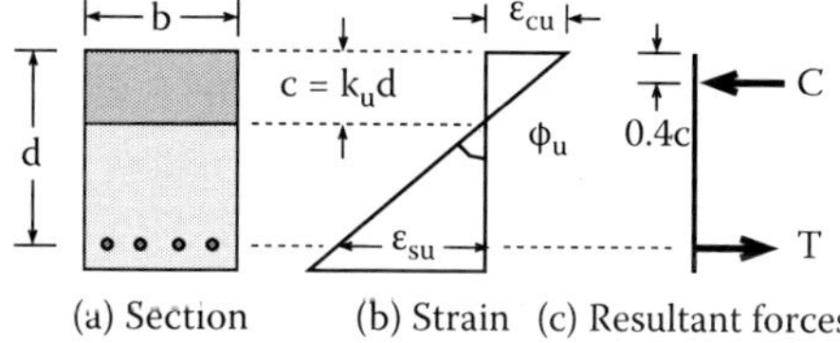

FIGURE 14.4 Strain and resultant forces at useful limit.

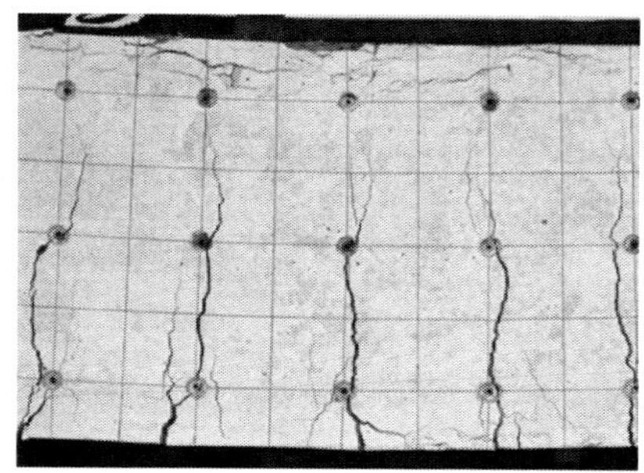

FIGURE 14.5 Failure of concrete in compression (top face of beam).

If the calculated strain in the reinforcement is more than f_y/E_s, where E_s is Young's modulus for the reinforcement, the moment and the unit curvature at point (3) are determined as

Moment = Tensile force in reinforcement × Internal lever arm

$$M_n \approx A_s \cdot f_y \cdot (1 - 0.4 \cdot k_u) \cdot d \tag{14.4}$$

The associated unit curvature is

$$\phi_u = \frac{\varepsilon_{cu}}{k_u \cdot d} \tag{14.5}$$

If the calculated value of ε_{su} exceeds 0.06, there may be a likelihood of fracture in the tensile reinforcement before the useful limit of concrete strain is reached. In that case, the limiting coordinates should be based on the strain and stress limits of the reinforcement. The reinforcement stress should be set equal to the strength of the reinforcement and the unit curvature should be determined from

$$\phi_u = \frac{\varepsilon_{su}}{d \cdot (1 - k_u)} \tag{14.6}$$

where ε_{su} is the limiting strain of reinforcement (Figure 4.2).

Because the strain in the concrete may be considerably lower than the limiting value, the value of k_u is better approximated in this case by using the expression for k given in Chapter 13. It would be instructive to obtain both k_u and k to observe the difference.

If desired or required, the limiting coordinates of the moment–curvature relationship may be determined using a more detailed procedure, as described below.

14.2 A DETAILED PROCEDURE FOR DETERMINING THE LIMIT TO THE MOMENT–CURVATURE RELATIONSHIP

Figure 14.3 shows how unit strains and stresses change as bending moment and imposed deformation increase. After yielding, it is assumed the stress in the reinforcement remains constant at f_y. And from steps 2 and 3, the shape of the distribution of unit stress in the concrete changes from a shape close to a triangle to a shape bounded by a parabola and an approximately straight line.

The equations describing the conditions of equilibrium (Chapter 11) are written in reference to the location and magnitude of stress resultants. Both location and magnitude are functions of the shape of the stress distribution, which changes with increase in concrete strain.

We follow the same procedure as we did earlier, but this time in reference to the stresses and strains depicted in Figure 14.6. The extreme fiber strain in concrete is commonly set at $\varepsilon_{cu} = 0.003$.

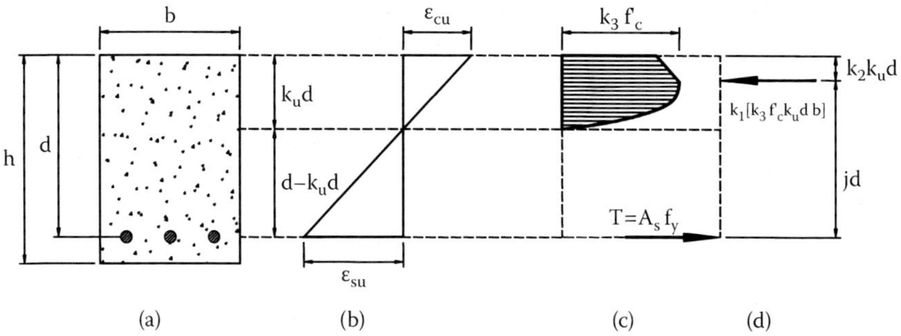

FIGURE 14.6 Assumed distributions of stress and strain at maximum moment. (a) Section. (b) Unit strain. (c) Concrete unit stresses. (d) Resultant.

As in the approximate procedure, we start by considering the equilibrium of internal forces assumed to be acting on the section. But in this derivation, we treat the development of the force in the concrete by referring to the assumed properties of the concrete (Figure 14.6). The compressive force C may be obtained by integrating the stresses from the neutral axis to the extreme fiber in compression. But the conventional procedure is to compute C using three coefficients (dimensionless) to define the properties of an "equivalent stress block":

k_1 = Ratio of the area within the compressive stress distribution to the area of an enclosing rectangle (Figure 14.7). This factor is assumed to be 0.85 for concretes having cylinder strengths of 4 ksi or less. For concretes with higher strengths, the approximate value of k_1 is defined by the exact expression

$$k_1 = 0.85 - 0.05 \cdot \left(f'_c - 4 \text{ ksi}\right) \geq 0.65 \tag{14.7}$$

k_2 = Ratio of the distance between the depth to the line of action of the compressive force and the neutral axis depth, assumed to be $k_1/2$ or 0.42 for concretes with strengths not exceeding 4 ksi.

k_3 = Ratio of the maximum compressive stress in the compression block to the strength determined from $6 \cdot 12$ in. test cylinders. This value is set at 0.85.

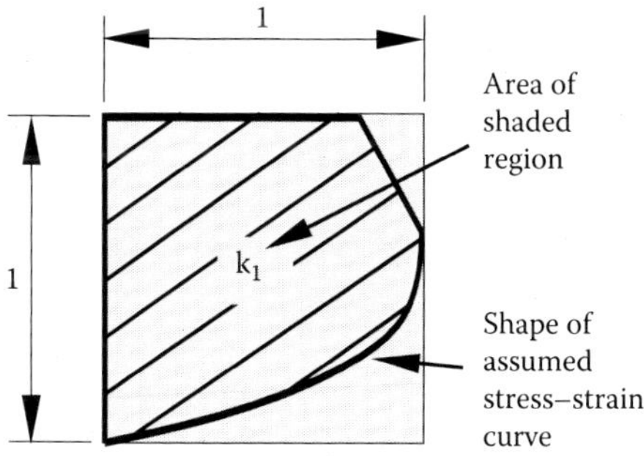

FIGURE 14.7 Definition of k_1.

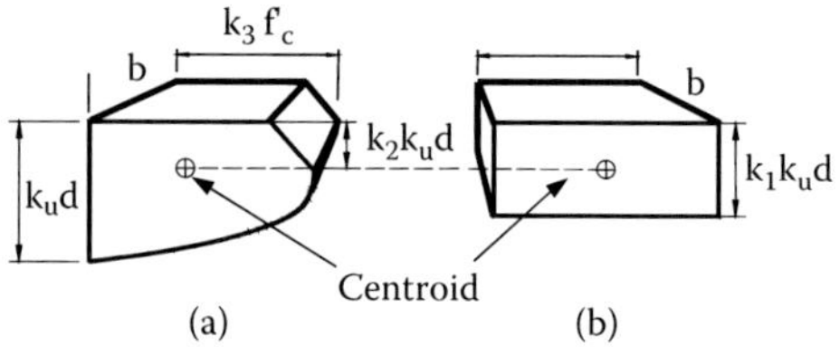

FIGURE 14.8 Equivalent stress block.

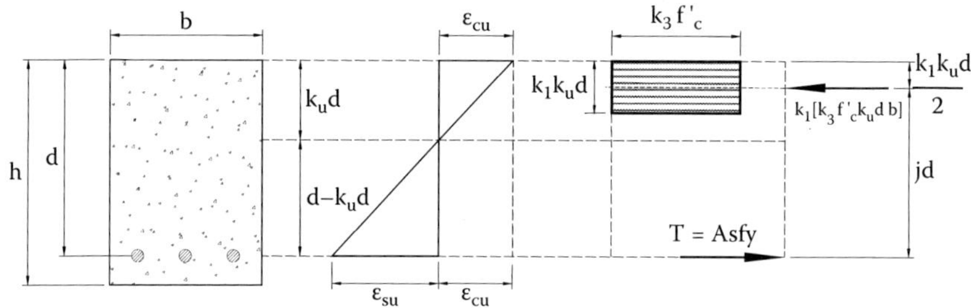

FIGURE 14.9 Idealized distributions of stress and strain at maximum moment.

Using the coefficients k_1 and k_3, we determine the resultant of compressive stresses in the concrete,

$$C = k_1 \cdot k_3 \cdot f'_c \cdot b \cdot k_u \cdot d \tag{14.8}$$

The result is the volume of the object shown in Figure 14.8a. Given the definition of k_1, this object has the same volume as the box shown in Figure 14.8b. Because k_2 is assumed to be $k_1/2$, the vertical distance from the top face to the centroid is the same for both objects. The stress distribution at maximum concrete strain can be idealized as shown in Figure 14.9.

We should keep in mind that the derivation was developed assuming that the width of the zone in compression is constant.

If the steel yields before the maximum strain in the concrete reaches 0.003, the force in the reinforcement is

$$T = \rho \cdot f_y \cdot b \cdot d \tag{14.9}$$

Equating the tensile and compressive forces acting on the section, ignoring tensile stresses in concrete, and solving for k_u, we obtain

$$k_u = \frac{\rho \cdot f_y}{k_1 \cdot k_3 \cdot f'_c} \tag{14.10}$$

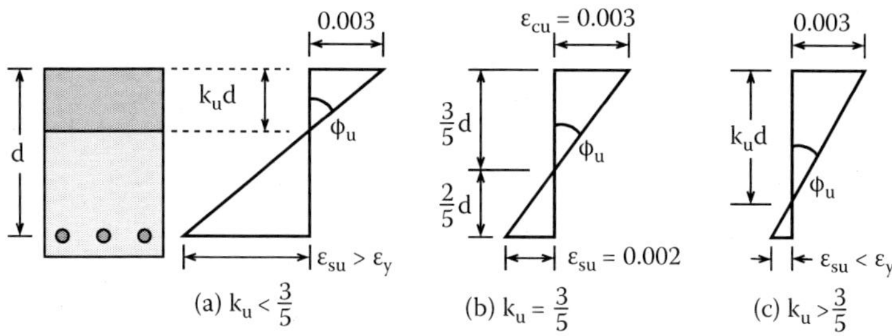

FIGURE 14.10 Distributions of strain with Grade 60 reinforcement.

Equation 14.10 gives us the relative depth to the neutral axis associated with a maximum strain in the concrete of $\varepsilon_{cu} = 0.003$. It is based on the assumption that the stress in the reinforcement is f_y. Figure 14.10b shows the strain distribution with $k_u = 3/5$ and $\varepsilon_{cu} = 0.003$, which leads to $\varepsilon_{su} = 0.002$. Recall that the yield strain of Grade 60 reinforcement is $\varepsilon_y = 60/29000 \sim 0.002$. For such a reinforcement, if $k_u < \sim 3/5$, then $\varepsilon_{su} > \varepsilon_y$ (Figure 14.10a). If $\varepsilon_{su} > \varepsilon_y$, we conclude $f_s = f_y$. If $\varepsilon_{su} < \varepsilon_y$ (Figure 14.10c), f_y should be replaced with $\varepsilon_{su} \cdot E_s$ in Equation 14.10 (which could then be used with Equation 14.3 to solve for k_u). If the limits to the reinforcement ratio that is defined in Chapter 16 are met (if the beam does not have too much reinforcement), the assumption $f_s = f_y$ is correct. In a beam with too much reinforcement—an overreinforced beam—the concrete tends to spall before the steel yields.

Taking moments about the centroid of the compressive force (Figure 14.9),

$$M_n = A_s \cdot f_y \cdot j \cdot d \tag{14.11}$$

with $j = (1 - k_2 \cdot k_u)$. For concrete with a compressive strength between 4 and 6 ksi, $f_y = 60$ ksi and $0.5\% < \rho < 2\%$. j varies between 0.82 and 0.97. Assuming j = 0.9 is satisfactory for initial proportioning.

Equation 14.11 defines nominal flexural capacity. It follows from a failure criterion based on a limiting value for the compressive strain in concrete. In conventional design, the nominal capacity is reduced by a strength reduction factor Φ to provide a margin of safety (design capacity = ΦM_n). The design capacity is not to be exceeded by demands computed for *factored* loads.

EXAMPLE 14.1

Draw an approximate moment-unit curvature diagram for the cross section shown in Figure 14.11. Assume the compressive strength of the concrete is 4000 psi.

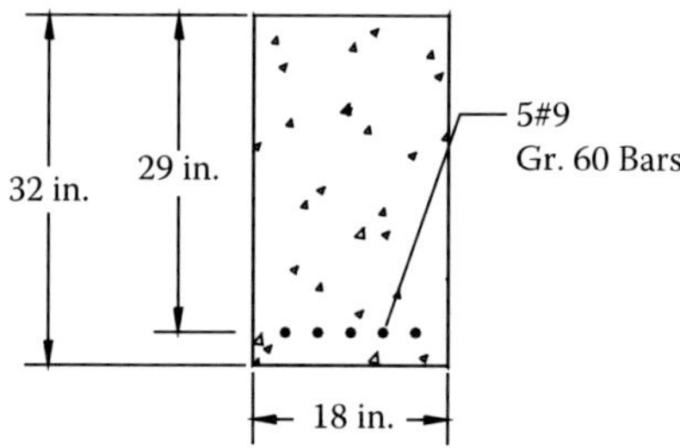

FIGURE 14.11 Beam section for Example 14.1.

SOLUTION

Coordinate Set (1): Cracking

Modulus of rupture (corrected to two significant figures): $f_r = 6 \cdot \sqrt{f'_c \cdot psi} = 380\ psi$.

Moment at cracking: $M_{cr} = \frac{b \cdot h^2}{6} \cdot f_r = 97\ kip \cdot ft$.

Unit curvature at cracking:

Young's modulus for concrete: $E_c = 57{,}000\sqrt{f'_c \cdot psi} = 3.6 \times 10^3\ ksi$.

Assumed strain at cracking: $\varepsilon_{cr} = \frac{f_r}{E_c} = 1 \times 10^{-4}$.

Unit curvature at cracking: $\phi_{cr} = \frac{\varepsilon_{cr}}{\frac{1}{2} \cdot h} = 7 \times 10^{-6}\ \frac{1}{in.}$.

Coordinate Set (2): Yielding

Modular ratio: $n = \frac{E_s}{E_c} = 8.0$.

Reinforcement ratio: $\rho = \frac{A_s}{b \cdot d} = 1.0\%$.

Ratio of neutral axis depth to effective depth: $k = \sqrt{\rho^2 \cdot n^2 + 2 \cdot \rho \cdot n} - \rho \cdot n = 0.32$.

Yield moment estimate: $M_y = A_s \cdot f_y \cdot d \cdot \left(1 - \frac{k}{3}\right) = 647\ kip \cdot ft$.

Unit curvature at yield:

Yield strain: $\varepsilon_y = \frac{f_y}{E_s} = 2.1 \times 10^{-3}$.

Unit curvature: $\phi_y = \frac{\varepsilon_y}{d \cdot (1-k)} = 1.1 \times 10^{-4}\ \frac{1}{in.}$.

Coordinate Set (3): Assumed Limit

Ratio, depth of the neutral axis at the limit to the effective depth: $k_u = \frac{\rho \cdot f_y}{0.7 \cdot f'_c} = 0.2$.

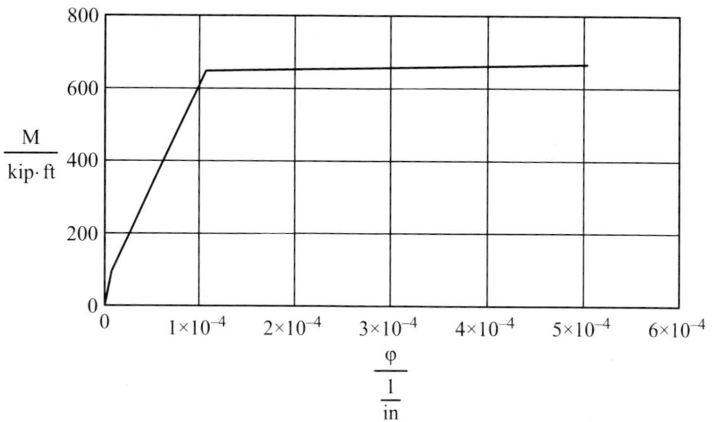

FIGURE 14.12 Moment–curvature relationship.

Check tensile reinforcement strain:

$$\varepsilon_{su} = \varepsilon_{cu} \cdot \frac{1-k_u}{k_u} = 0.012,$$

which is more than yield strain (as assumed) and does not exceed 0.06.

Limit moment: $M_n = A_s \cdot f_y \cdot d \cdot (1-0.4k_u) = 665$ kip · ft .

Limit curvature: $\phi_u = \dfrac{\varepsilon_{cu}}{k_u \cdot d} = 5 \times 10^{-4} \dfrac{1}{\text{in.}}$.

Figure 14.12 shows the moment–curvature relationship connecting coordinate sets 1, 2, and 3.

Detailed Calculations for Coordinate Set 3

We use $k_1 = 0.85$; $k_2 = 0.42$; $k_3 = 0.85$.

Ratio, depth of the neutral axis at the limit to the effective depth: $k_u = \dfrac{\rho \cdot f_y}{k_1 \cdot k_3 \cdot f_c'} = 0.20$.

Check tensile reinforcement strain: $\varepsilon_{su} = \varepsilon_{cu} \cdot \dfrac{1-k_u}{k_u} = 0.012$.

Limit moment: $M_n = A_s \cdot f_y \cdot (1-k_2 \cdot k_u) \cdot d = 664$ kip · ft .

Limit curvature: $\phi_u = \dfrac{\varepsilon_{cu}}{k_u \cdot d} = 5 \times 10^{-4} \dfrac{1}{\text{in.}}$.

We find the results of the detailed procedure to differ negligibly from those of the approximate procedure. Next, we repeat the entire process for a section with twice the amount of reinforcement to see if there will be a difference in that case.

EXAMPLE 14.2

Draw an approximate moment-unit curvature diagram for the cross section shown in Figure 14.13. Assume the compressive strength of the concrete is 4000 psi.

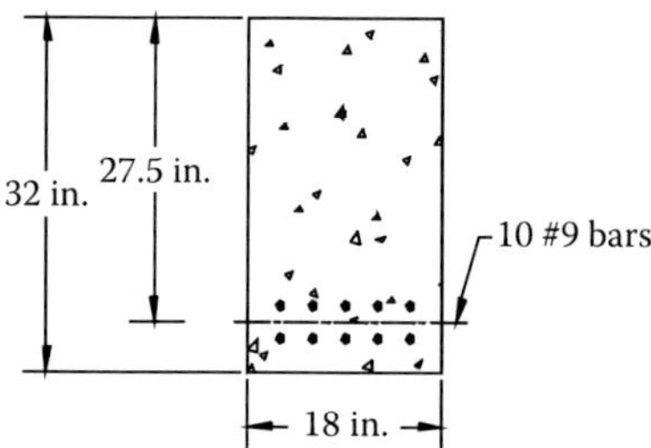

FIGURE 14.13 Beam section for Example 14.2.

SOLUTION

Coordinate Set (1): Cracking

The first set of coordinates is the same as in the previous example:

Moment at cracking: $M_{cr} = \dfrac{b \cdot h^2}{6} \cdot f_r = 97 \text{ kip} \cdot \text{ft}.$

Unit curvature at cracking: $\phi_{cr} = \dfrac{\varepsilon_{cr}}{\frac{1}{2} \cdot h} = 7 \times 10^{-6} \dfrac{1}{\text{in.}}.$

Coordinate Set (2): Yielding

Modular ratio: $n = \dfrac{E_s}{E_c} = 8.0.$

Reinforcement ratio: $\rho = \dfrac{A_s}{b \cdot d} = 2.0\%.$

Ratio of the neutral axis depth to the effective depth:

$$k = \sqrt{\rho^2 \cdot n^2 + 2 \cdot \rho \cdot n} - \rho \cdot n = 0.43.$$

Yield moment estimate: $M_y = A_s \cdot f_y \cdot d \cdot \left(1 - \dfrac{k}{3}\right) = 1.2 \times 10^3 \text{ kip} \cdot \text{ft}.$

Unit curvature at yield: $\phi_y = \dfrac{\varepsilon_y}{d \cdot (1-k)} = 1.3 \times 10^{-4} \dfrac{1}{\text{in.}}.$

Coordinate Set (3): Assumed Limit

Ratio, depth of the neutral axis at the limit to the effective depth: $k_u = \dfrac{\rho \cdot f_y}{0.7 \cdot f'_c} = 0.4.$

Check tensile reinforcement strain:

$$\varepsilon_{su} = \varepsilon_{cu} \cdot \frac{1-k_u}{k_u} = 3.9 \times 10^{-3},$$

which is more than yield strain (as assumed), does not exceed 0.06, and is much smaller than in the previous example.

Limit moment: $M_n = A_s \cdot f_y \cdot d \cdot (1 - 0.4k_u) = 1.1 \times 10^3 \text{ kip} \cdot \text{ft}.$

Limit curvature: $\phi_u = \dfrac{\varepsilon_{cu}}{k_u \cdot d} = 2.5 \times 10^{-4} \dfrac{1}{\text{in.}}.$

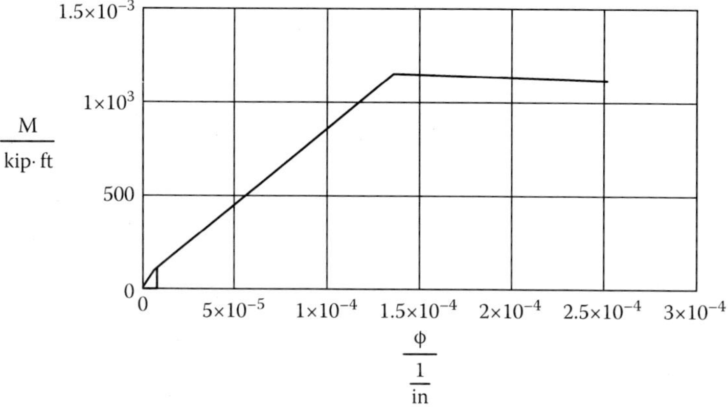

FIGURE 14.14 Moment–curvature relationship.

Figure 14.14 shows the moment–curvature relationship connecting coordinate sets 1, 2, and 3.

The calculated M_n is lower than the calculated M_y. That is unusual and has occurred because in developing the expression for yield moment it was assumed that concrete was in its linear stage of response.

In this case, the calculated concrete strain at yield is $\varepsilon_c = \phi_y \cdot k \cdot d = 1.6 \times 10^{-3}$. The calculated maximum stress in the concrete is $f_c = E_c \cdot \varepsilon_c = 5.6 \times 10^3$ psi. The calculated stress f_c exceeds $0.5f_c'$ by a large margin, indicating that the yield moment is overestimated by the expression used for M_y. In the following section a different method is presented to estimate M_y.

Detailed Calculations for Coordinate Set 3

We use $k_1 = 0.85$; $k_2 = 0.42$; $k_3 = 0.85$.

Ratio, depth of the neutral axis at the limit to the effective depth: $k_u = \dfrac{\rho \cdot f_y}{k_1 \cdot k_3 \cdot f_c'} = 0.4$.

Check tensile reinforcement strain: $\varepsilon_{su} = \varepsilon_{cu} \cdot \dfrac{1 - k_u}{k_u} = 4 \times 10^{-3}$.

Limit moment: $M_n = A_s \cdot f_y \cdot (1 - k_2 \cdot k_u) \cdot d = 1.1 \times 10^3 \text{ kip} \cdot \text{ft}$.

Limit curvature: $\phi_u = \dfrac{\varepsilon_{cu}}{k_u \cdot d} = 2.6 \times 10^{-4} \dfrac{1}{\text{in.}}$.

We find the results of the detailed procedure to differ by less than 3% from those of the approximate procedure if ρ = 2%.

EXAMPLE 14.3

For the beam shown in Figure 14.15 and with the properties listed below, select reinforcement to satisfy flexural strength requirements limiting the maximum

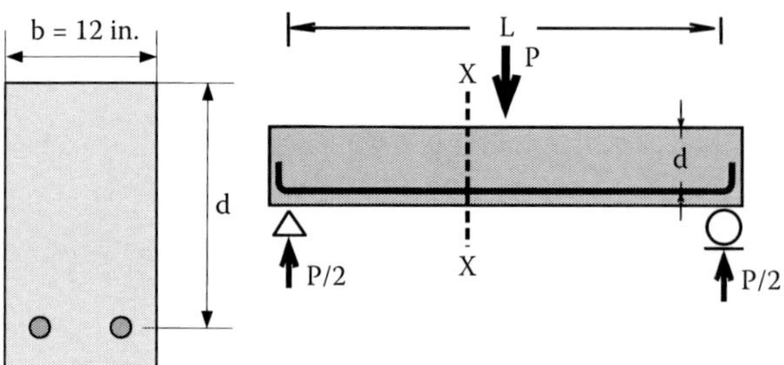

FIGURE 14.15 Beam for Example 14.3.

strain in the concrete to 0.003. Use a load factor of 1.4 and a strength reduction factor $\Phi = 0.9$ (Chapter 3).

PROPERTIES

Span L = 30 ft, service load P = 35 kip, $f'_c = 3$ksi, $f'_y = 60$ ksi, effective depth d = 25 in. Ignore self weight.

SOLUTION

1. Compute moment demand: M_u = Load factor × P × L/4 = 1.4 × 35 kip × 30 ft/4 = 368 kip-ft.
2. Assume j = 0.9 and $f_{su} = f_y$ and compute

$$A_s = M_u/(\Phi \text{ jd } f_y) = 368 \text{ kip-ft} \times 12 \text{ in./ft}/(0.9 \times 0.9 \times 25 \text{ in.} \times 60 \text{ ksi}) = 3.6 \text{ in.}^2$$

3. Choose an integer number of standard bars of the same diameter: four #9 bars (A_s = 4 in.^{2}).
4. Compute $\rho = A_s/(bd) = 4$ in.^{2}/(12 in. × 25 in.) = 1.33%.
5. Compute $k_u = \rho \cdot f_y/(k_1\ 0.85f'_c) = 0.0133 \times 60$ ksi/(0.85 × 0.85 × 3 ksi) = 0.37 (<3/5; therefore, $f_{su} = f_y$).
6. Compute $j = 1 - k_1\ k_u/2 = 1 - 0.85 \times 0.37/2 = 0.84$ (the computed value of j does not always match the assumed value of 0.9).
7. Compute nominal capacity $A_s \cdot f_y \cdot j \cdot d$ = 4 in.^{2} × 60 ksi × 0.84 × 25 in. × 1 ft/12 in. = ~420 kip-ft.
8. Compute the ratio of design capacity to demand $\Phi A_s \cdot f_y \cdot jd/M_u$ = 0.9 × 420 kip-ft/368 = ~1.0, OK.

We conclude that four Grade 60 #9 reinforcing bars would provide adequate reinforcement.

BARE ESSENTIALS

If $\rho < 2\%$,

$$M_n \approx A_s \cdot f_y \cdot 0.9 \cdot d$$

EXERCISES

1. Repeat the example for L = 24 ft.
2. For beam segment AB in the example from Chapter 3, select reinforcement to be provided near each support and near mid-span. Assume f'_c = 5000 psi, f_y = 60 ksi, and a strength reduction factor of $\Phi = 0.9$. Do not ignore self weight.

15 Development of a Quantitative Relationship between Moment and Unit Curvature

In Chapters 12 and 13 we learned how to relate moment and curvature for uncracked and cracked rectangular sections reinforced in tension. But we dealt exclusively with cases in which the materials were assumed to remain linear. In Chapter 14 we learned how to determine the flexural strength and limiting unit curvature of rectangular sections reinforced in tension. In relation to Figure 11.5, we have dealt with Stages I and II, and with the end (the limit) of the curve representing the relationship between moment and curvature. But we have not dealt with intermediate cases. In this chapter, we learn how to construct the entire relationship of moment with unit curvature from the beginning to the limiting condition. The curve that defines the variation of moment with change in unit curvature for a beam section is a very important property of the section because it shows in one two-dimensional plot the strength of the section, its initial stiffness, its ability to remain integral after reaching its strength, and its toughness or energy-absorbing capacity. Even though it is seldom required in a routine design task, it is essential for a designer to be able to construct it and be familiar with its shape, as well as understand the variables to which it is sensitive.

In this chapter, we shall construct such a curve for a specific cross section shown in Figure 15.1, assuming the properties listed below for its materials and dimension.

EXAMPLE 15.1

MATERIAL PROPERTIES AND DIMENSIONS

Concrete cylinder compressive strength	$f'_c = 4200$ psi
Young's modulus of concrete	$E_c = \sqrt{f'_c} = 3.7 \times 10^3$ ksi
Yield stress of concrete	$f_y = 60$ ksi
Young's modulus of reinforcing steel	$E_s = 29{,}000$ ksi
Amount of tensile reinforcement	$A_s = 4$ in.
Width of cross section	$b = 16$ in.
Effective depth of cross section	$d = 25$ in.

Continued

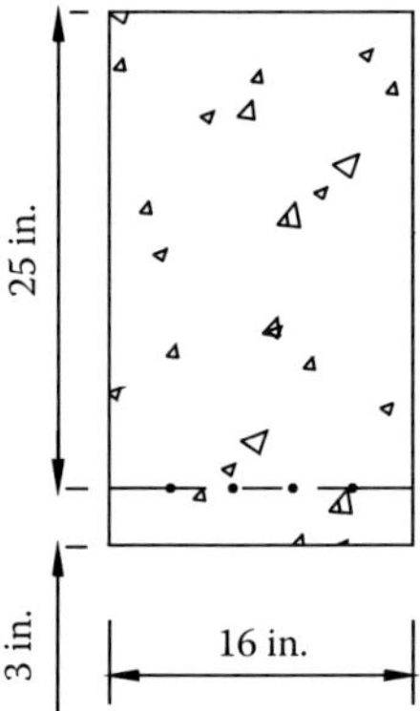

FIGURE 15.1 Concrete stress–strain curve.

Height of section	h = 28 in.
Tensile reinforcement ratio	$\rho = \frac{A_s}{bd} = 1\%$
Modulus of rupture (corrected to two significant figures)	$f_r = 6\sqrt{f'_c} = \sim 390$ psi

STRESS–STRAIN RELATIONSHIP OF REINFORCING STEEL

The stress–strain relationship of the reinforcing steel is assumed to be elasto-plastic, as determined below and shown in Figure 14.2. The expression for the stress–strain curve of the reinforcement (Equation 15.1) states that the steel stress increases linearly with strain ($E_s \times \varepsilon_s$) up to the yield strain $\varepsilon_y = f_y/E_s$. After that limit is reached, the steel stress remains constant as strain increases.

$$f_s = E_s \times \varepsilon_s \le f_y \tag{15.1}$$

STRESS–STRAIN RELATIONSHIP OF REINFORCING CONCRETE

The assumed stress–strain relationship for concrete in compression is defined by Equations 5.11 and 5.12, which require two parameters:

Concrete compressive strength (assumed to be 85% of cylinder strength, Chapter 7)	$f''_c = 0.85 \times 4200$ psi
Assumed strain at maximum compressive stress	$\varepsilon_o = 2\ f''_c \div E_c = 0.002$

The curve defined by Equations 5.11 and 5.12 using the values of f''_c and ε_o is plotted in Figure 15.2.

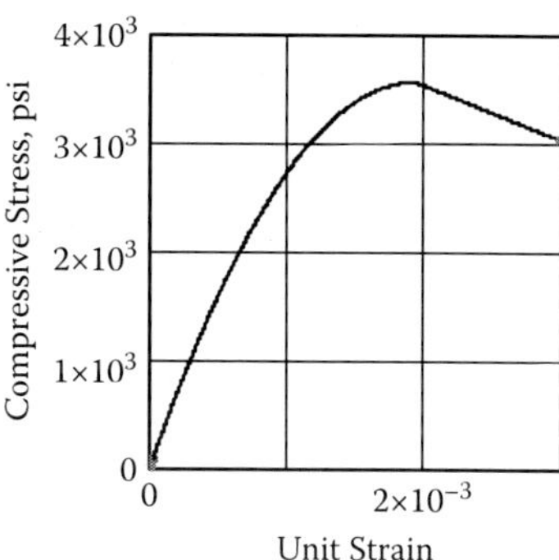

FIGURE 15.2 Cross section of beam.

FIRST STEP

The first step involves the determination of the moment and unit curvature coordinates at cracking. Considering the expected scatter in tensile strength (Figure 5.6), we understand that these coordinates are approximate, but we compute them if only to get a measure of the initial slope of the curve.

From Equation 12.7:

Cracking moment $$M_{cr} = \frac{bh^2}{6} f_r = 68 \text{ kip} \cdot \text{ft}$$

From Equation 12.9:

Curvature at cracking: $$\varphi_{cr} = \frac{f_r}{E_c \cdot \frac{h}{2}} = 8 \times 10^{-6} \frac{1}{\text{in.}}$$

For the remaining part of the moment–curvature relationship, we ignore the contribution of concrete in tension.

REMAINING STEPS

We develop the curve one point at a time. Determining each point requires eight steps:

1. A small (but not smaller than $f_r \div E_c$) compressive strain is assumed in the extreme fiber in compression.
2. In reference to Figure 15.3 we know that the average compressive stress over the area extending from the neutral axis to the extreme fiber in compression will be the same no matter how deep the neutral axis is. We determine the average stress corresponding to the chosen maximum strain.
3. Using the computed average compressive stress and the given magnitude of the reinforcement ratio, we determine and plot the relationship between steel strain and steel stress required to satisfy the condition of equilibrium of axial forces (C = T in Figure 15.3).

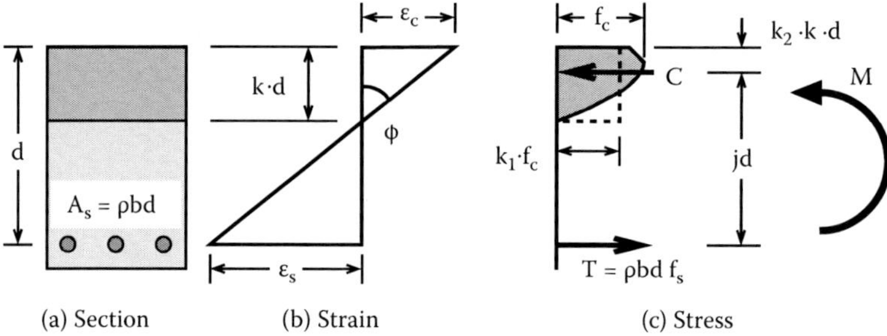

FIGURE 15.3 Cross section of beam.

4. We superimpose the stress–strain curve for the reinforcement (Equation 15.1) on the plot developed in step 3.
5. The intersection point of the two curves gives us the steel strain and stress compatible with the compression strain assumed in step 1.
6. The steel strain determined in step 5 helps us determine the depth of the neutral axis ($k \cdot d$ in Figure 15.3). The unit curvature ϕ is obtained as the ratio of the assumed strain at the extreme fiber in compression (step 1) to the depth of the neutral axis.
7. We determine the location of the centroid of the compressive force on the section (distance $k_2 \cdot k \cdot d$ in Figure 15.3).
8. The moment capacity is determined as the product of the force in the reinforcement (T) and the distance between the centroids of compressive and tensile forces on the section ($jd = d - k_2 \cdot k \cdot d$).

Step 8 completes the procedure to determine the moment and unit curvature coordinates for the selected strain in the extreme fiber in compression. Its results are saved and the steps are repeated for a selected higher value of the maximum compressive strain until the limiting strain in compression is reached.

SOLUTION

Step 1: Let us start by assuming the unit strain in the extreme fiber in compression to be $\varepsilon_c = 0.0003$.

Step 2: The mean compressive stress is obtained from Equation 15.2. The procedure is simply determining the area under the stress–strain curve up to a selected strain ε_c by integrating the curve in Figure 15.2 and dividing the result by the base of the area.

$$f_{c\ average} = \frac{1}{\varepsilon_c}\int_0^{\varepsilon_c} f_c(\varepsilon)d\varepsilon = \int_0^1 f_c(\alpha\varepsilon_c)d\alpha = \sim 530\ \text{psi} \tag{15.2}$$

where $\alpha = \varepsilon/\varepsilon_c$.

Step 3: Equilibrium of forces in the axial direction requires

$$C = f_{c\ average} k \cdot d\, b = A_s f_s = T \tag{15.3}$$

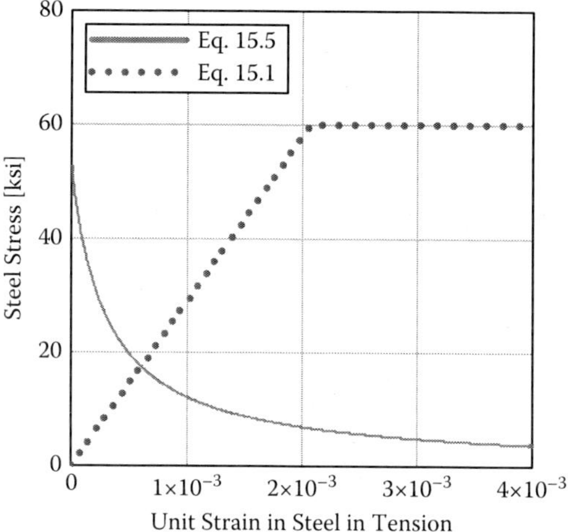

FIGURE 15.4 Cross section of beam.

From Figure 15.3:

$$k^* = \varepsilon_c \div (\varepsilon_c + \varepsilon_s) \tag{15.4}$$

Combining these two equations and solving for f_s:

$$f_s = \frac{f_{c\ \text{average}}}{\rho} \frac{\varepsilon_c}{(\varepsilon_c + \varepsilon_s)} \tag{15.5}$$

This expression is plotted in Figure 15.4.

Step 4. The stress–strain relationship given by Equation 15.1 is also plotted in Figure 15.4.

Step 5: The intersection of the curves plotted in Figure 15.4 is determined to be approximately $\varepsilon_s = 0.0006$. This result can be obtained by zooming in on the figure, by trial and error, or by the numerical algorithms commonly called successive approximations. Learning the method for successive approximations proposed by Isaac Newton to find the root of a function (the difference between Equations 15.5 and 15.1 in this case) is particularly interesting because it helps the engineer appreciate Newton's genius.

The unit stress at the intersection of the two curves is $f_s = 17.5$ ksi.

Step 6: The depth to the neutral axis from the extreme fiber in compression is determined from

$$k^* = \varepsilon_c \div (\varepsilon_c + \varepsilon_s) = 0.0003 \div (0.0003 + 0.0006) = 1/3$$

The curvature associated with the concrete strain selected is

$$\varphi = \varepsilon_c \div (k \cdot d) = 0.0003 \div (25\ \text{in.}/3) = 3.6 \times 10^{-5}\ \tfrac{1}{\text{in.}} \tag{15.6}$$

Step 7: The distance from the extreme fiber in compression to the centroid of the compressive stress distribution is determined from Equation 15.7:

$$k_2^* = \frac{\int_0^1 f_c(\alpha\varepsilon_c)(1-\alpha)\,d\alpha}{\int_0^1 f_c(\alpha\varepsilon_c)\,d\alpha} = 0.34 \tag{15.7}$$

The numerator of Equation 15.7 represents the first moment of the nonlinear stress distribution about the outermost fiber in compression. The denominator represents the area of the stress distribution.

For a small strain in the extreme fiber in compression, k_2^* is very close to what we would determine for a linear variation of stress (1/3), and the value obtained from Equation 15.7 does not surprise us.

Step 8: The moment associated with $\varepsilon_c = 0.0003$ is

$$M = A_s f_s (1 - k \cdot k_2) d = 129 \text{ kip ft} \tag{15.8}$$

We repeat the above set of steps for increasing values of ε_c to obtain the moment–curvature relationship shown in Figure 15.5.

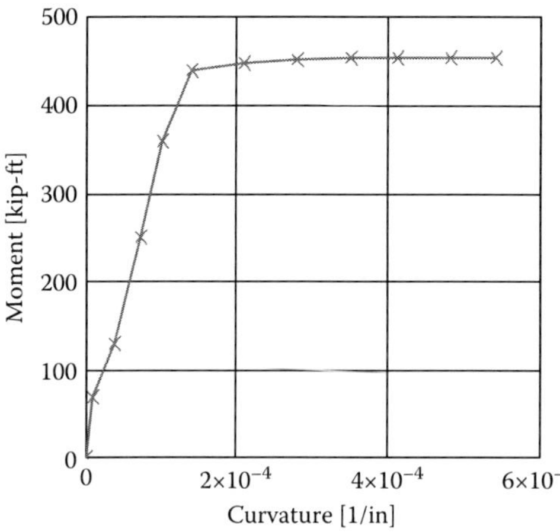

Moment (kip-ft)	Curvature (1/in.)
68	0.000008
129	0.000036
249	0.000071
358	0.000100
439	0.000140
447	0.000210
451	0.000280
454	0.000350
454	0.000410
454	0.000480
454	0.000540

FIGURE 15.5 Cross section of beam.

BARE ESSENTIALS

The moment–curvature relationship for a moderately reinforced section ($\rho < 2\%$) resembles qualitatively the stress–strain relationship of the tensile reinforcement.

EXERCISE

Plot moment–curvature diagrams for the beam described in Example 15.1 using

1. $A_s = 8$ in.^{2}
2. $A_s = 12$ in.^{2}

Superimpose your results and those in Figure 15.5. In a single sentence, what do you conclude from this exercise?

16 Maximum and Minimum Amounts of Longitudinal Reinforcement for Beams

The amount of reinforcement required at a beam section is determined by calculation in reference to the moment demand and the material and dimensional properties of the section, but there are limits to the amount of longitudinal reinforcement set primarily by accumulated experience of the profession. The minimum ratio of reinforcement at a section is governed by a behavioral concern, and the maximum amount is governed by two issues: one related to constructability and the other to behavior.

The constructability issue that limits the maximum amount of reinforcement has to do primarily with casting of concrete. There is a minimum tolerable distance between the individual bars and between their layers that limits the maximum amount of reinforcement. This maximum is usually expressed in terms of the reinforcement ratio. Generally, it is set close to $\rho \le 0.02$, applies to compressive as well as tensile reinforcement, and may vary depending on local tradition.

The minimum amount of tensile reinforcement at a section is said to be controlled by the desire to avoid a flexural failure initiated by fracture of reinforcement because such a failure is likely to occur with no warning. To determine the minimum amount of reinforcement by calculation is not easy. The likelihood of fracture depends on many variables with large scatter, such as the fracture strain of reinforcement, bond between bar and concrete, and tensile strength of concrete. In U.S. construction, the minimum for beams was once set at $\rho = 0.005$. In the 1960s, when use of reinforcement with yield stress higher than 60 ksi became popular, a compromise was made on the debatable but convenient assumption* that the minimum reinforcement ratio for tensile reinforcement should be determined from the condition:

$$\text{Moment capacity of section} \ge \text{Cracking moment of section}$$

* There was nothing explicit in the traditions of reinforced concrete that the question of minimum reinforcement had been selected in reference to the cracking moment. Had minimum reinforcement been seriously related to flexural cracking moment, a quantity difficult to determine because of the variability of tensile strength and effects of shrinkage and temperature, then the relevant criterion should also involve factors such as shape of section and type of aggregate. The driving desire to reduce the amount of minimum reinforcement for bars with yield stress higher than 60 ksi was introduced in the 1960s. The desired reduction was achieved by relating the yield moment to the assumed cracking moment. Undeniably, the result is reasonable, but also undeniably, it could have been reached in a different manner without invoking the elusive magnitude of the cracking moment.

For a beam with a minimum ratio of reinforcement (ρ_{min}),

$$\text{Moment capacity} \cong 0.9 \cdot \rho_{min} \cdot b \cdot d^2 \cdot f_y \tag{16.1}$$

If we assume that the mean modulus of rupture is $8\sqrt{f'_c}$ (which is certainly plausible given the data in Figure 5.6) and the effective depth d is 0.7 × the height of the section,

$$\text{Cracking moment} = \frac{1}{6} b \cdot h^2 \cdot f_r = \frac{1}{6} b \cdot \left(\frac{d}{0.7}\right)^2 \cdot 8\sqrt{f'_c} \tag{16.2}$$

Equating Equations 16.1 and 16.2, we obtain

$$\rho_{min} \cong 3\frac{\sqrt{f'_c}}{f_y} \tag{16.3}$$

If we substitute the typical material strengths used for concrete in the 1930s, we obtain approximately $\rho = 0.005$, and that gives us the rationalization, questionable at best, to use it as a vehicle for current construction with higher-strength materials, provided we do not go below a lower bound. A reasonable value for the lower bound is $\rho = 0.002$.

Figure 16.1 compares the moment–curvature relationships of beam sections with reinforcement ratios of 1 and 0.1%, assuming $f'_c = 5000$ psi, $f_y = 60$ ksi, and d/h = 5/6. The tensile strength of the concrete was assumed to be $f_t = 10\sqrt{f'_c} = 10 \cdot \sqrt{5000}\text{psi} =$ 710 psi. Earlier, we introduced a reasonable lower bound for a tensile strength of $6\sqrt{f'_c}$. We use higher values here intentionally. The phenomenon we described in reference to the minimum amount of reinforcement required is one in which increases in the tensile strength of the concrete can be seen as having a negative effect. The curve shown for $\rho = 0.1\%$ illustrates a case in which the moment capacity, expressed as $M_n/(bd^2) = \rho \cdot j \cdot f_y \approx 0.001 \times 0.9 \times 60$ ksi $= 0.054$ ksi, is less than the moment at cracking M_{cr}, which we normalized as $M_n/(bd^2) = 0.17$ ksi. A similar beam would fail abruptly under gravity loads if the applied moment reaches the cracking moment.

The maximum reinforcement ratio related to behavior ρ_{max} has its roots in a statement by A. N. Talbot (1904):

> The action during the last stage [of behavior] differs for the two classes of beams (a) beams which may be termed normal beams in which the amount of reinforcement is not sufficient to develop the full compressive strength of the concrete at the maximum load and (b) beams which fail by crushing of the concrete before the elastic limit of the steel is reached and hence may be said to have an excess of [steel].

Currently, the first group cited (class a) by Talbot is classified as underreinforced beams, and the second group is classified as overreinforced beams. Building codes have discouraged the use of overreinforced sections because the flexural strength of such sections is sensitive to that of concrete and because they sustain brittle failures.

We estimate the reinforcement threshold separating underreinforced and overreinforced beams by studying the condition in which the limiting unit strain in concrete

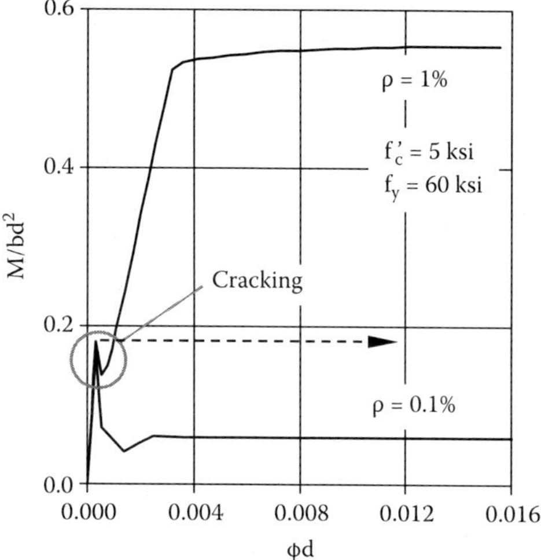

FIGURE 16.1 Moment–curvature relationships of beam sections with reinforcement ratios of 1% and 0.1%.

and the unit strain at yield of steel are reached simultaneously. This condition is called balanced failure. The term *balanced* can be traced back to Talbot (1907) too:

> The writer calls "the balanced reinforcement" ... the amount of reinforcement for which the elastic tensile strength of the steel used for reinforcement may be considered to balance the compressive strength in the concrete of the beam....
>
> The term "balanced reinforcement" referred to above is a convenient term for general use. It should be taken to mean that amount of reinforcement for which the allowable stress in the steel and the allowable stress in the concrete both exist at the same time. The determination of the balanced reinforcement for given conditions of materials, fabrication, and use is a matter involving calculation and experimentation, but in any event the judgment of the designer must enter into the choice of the amount.

The distribution of unit strain at balanced failure is depicted by the thick solid line in Figure 16.2.

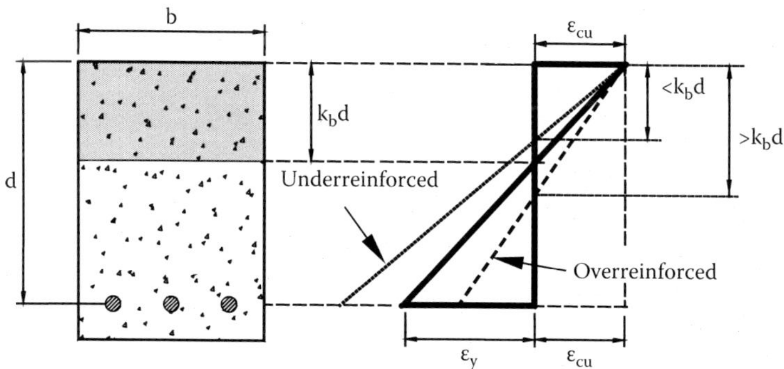

FIGURE 16.2 Distribution of unit strain at balanced failure.

From strain geometry:

$$k_b \cong \frac{\varepsilon_{cu}}{\varepsilon_y + \varepsilon_{cu}} \tag{16.4}$$

From equilibrium of normal forces:

$$\rho_b \cong \frac{k_1 k_3 f_c'}{f_y} k_b \tag{16.5}$$

where k_b is the ratio of the neutral axis depth to the effective depth at simultaneous development of the useful limit of compressive strain in the concrete and the yield strain in the tensile reinforcement, ε_{cu} is the useful limit of strain in compressed concrete, ε_y is the yield strain of tensile reinforcement, k_1 is the ratio of area of a compressive stress block profile to that of an enclosing rectangle, k_3 is the ratio of the concrete strength of a beam to that determined from test cylinders, f_c' is the compressive strength of concrete determined from tests of 6 × 12 in. cylinders, and f_y is the yield stress of tensile reinforcement.

The accuracy of ρ_b may be judged considering that it is acceptable to assume ε_{cu} to be 0.003 or 0.004 in the study of the behavior of reinforced concrete beams. To develop a sense of proportion, we assume $k_1k_3 = 0.7$, $\varepsilon_{cu} = 0.003$, $\varepsilon_y = 0.002$, $f_c' =$ 4000 psi, and $f_y = 60{,}000$, and find that the corresponding balanced reinforcement ratio is approximately

$$\rho_b \cong \frac{0.7 \cdot 4000 \cdot \text{ksi}}{60000 \cdot \text{ksi}} \cdot \frac{0.003}{0.003 + 0.002} = \frac{0.7 \cdot 4000 \cdot \text{ksi}}{60000 \cdot \text{ksi}} \cdot \frac{3}{5} = \sim 0.03$$

which suggests that it takes a reinforcement ratio of approximately 3% to overreinforce a rectangular section. Knowing the possible scatter in concrete strength, in the assumed strains, and the linearity of strain, we infer that to make certain a section is underreinforced, a reinforcement ratio of no more than 2% ought to be used in beams. Current standards in many countries limit the maximum reinforcement ratio to approximately ¾ of ρ_b. As will be described in Chapter 17, larger amounts of reinforcement may be used if the section has compression reinforcement.

EXAMPLE 16.1

A simple beam spanning 30 ft is to be proportioned using the parameters given and for two cases:

1. d = 25 in.
2. d = 16 in.

Select a longitudinal reinforcement to satisfy flexural strength requirements ($\Phi M_n > M_u$). Express the required amount of steel as a reinforcement ratio and compare it with ρ_b and $\rho_{max} = ¾\ \rho_b$. Use a load factor of 1.4.

Ignore self weight.

SOLUTION

1. d = 25 in. (Figure 16.2). We solved this case in Chapter 14. We concluded that four #9s would suffice. The corresponding ratio of reinforcement is 1.3%. This ratio is smaller than

$$\rho_b = \frac{3}{5} \cdot \frac{k_1 \cdot 0.85 \cdot f_c'}{f_y} = \frac{3}{5} \frac{0.85 \times 0.85 \times 3 \text{ ksi}}{60 \text{ ksi}} = 2.2\%$$

and

$$\rho_{max} = \frac{3}{4} \cdot \rho_b = 1.6\%$$

2. d = 16 in. (Figure 16.3). We start by assuming that j = 0.9 and $f_s = f_y$:

$$A_s = \frac{1.4 \times 35 \text{ kip} \times \frac{1}{4} \times 30 \text{ ft} \times 12 \frac{\text{in.}}{\text{ft}}}{0.9 \times 60 \text{ ksi} \times j \times 16 \text{ in.}} \approx 5.7 \text{ in.}^2$$

We choose six #9s, realizing that we may need to arrange them in two layers to fit into the space we have (Figure 16.4). We define d as the distance to the centroid of the reinforcement.

$$\rho = \frac{6 \text{ in.}^2}{16 \text{ in.} \times 12 \text{ in.}} = 3.1\%$$

We notice that this reinforcement ratio is not desirable because it exceeds ρ_{max} = 1.6%. At this point one should reconsider the proportions of the beam or the materials selected.

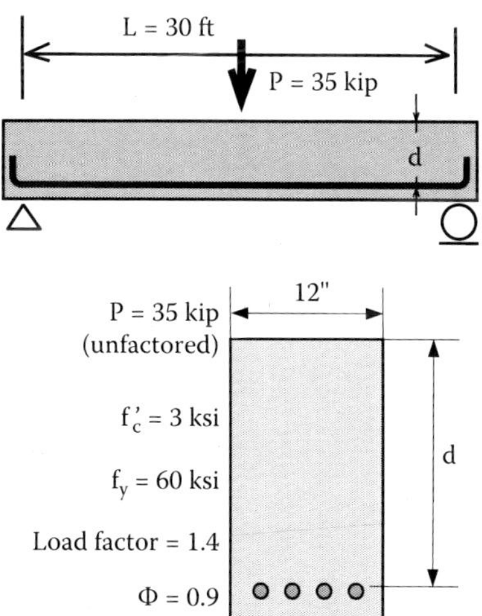

FIGURE 16.3 Example 16.1, case 2.

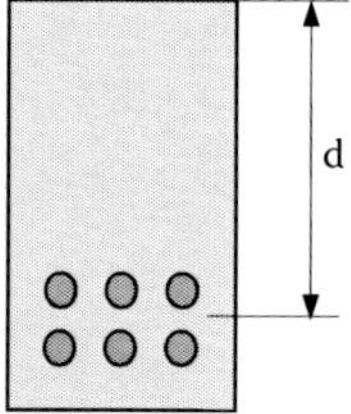

FIGURE 16.4 #9 bars arranged in two layers.

Nevertheless, we proceed to compute the nominal flexural capacity of the beam for the chosen area of steel because it is instructive to see how conditions vary for high reinforcement ratios.

We compute k_u assuming $f_s = f_y$:

$$k_u = \frac{3.1\% \times 60 \text{ ksi}}{0.85 \times 0.85 \times 3 \text{ ksi}} = 0.86 \qquad > 0.6 \Rightarrow f_s < f_y$$

Because $k_u > 3/5$, we conclude that the steel does not yield before the concrete reaches its useful limit of strain. We therefore need to revise our estimate of the stress in the reinforcement. We do so using the graphical solution we used in Chapter 15 (Figure 16.4). We compare the functions:

$$f_s(\varepsilon_{su}) = \frac{\varepsilon_{cu}}{\varepsilon_{cu} + \varepsilon_{su}} \cdot \frac{k_1 \cdot 0.85 \cdot f_c'}{\rho} \quad \text{and} \quad f_s(\varepsilon_{su}) = E_s \cdot \varepsilon_{su} \le f_y$$

and obtain $f_s = 45.5$ksi (Figure 16.5). With this value we revise k_u:

$$k_u = \frac{3.1\% \times 45.5 \text{ ksi}}{0.85 \times 0.85 \times 3 \text{ ksi}} = 0.65$$

We get $k_u > 3/5$ again. But that is now consistent with the estimate $f_s = 45.5$ ksi $< f_y$.

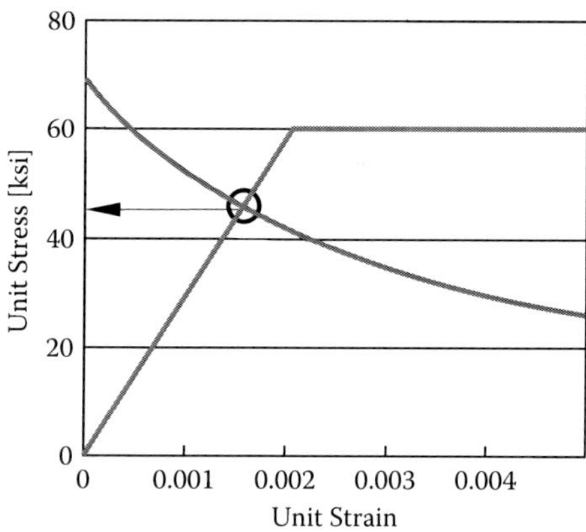

FIGURE 16.5 Graphical solution.

Now the internal arm is given by j = 1−(0.85/2) × 0.65 = 0.72.

This estimate deviates from the value of 0.9 (which we assumed) more than what we usually expect because we are working outside the ranges in which we observed j to be close to 0.9. We should remember that j is smaller for larger reinforcement ratios.

We now compute moment capacity:

$$\Phi M_n = 0.9\left(6 \text{ in.}^2 \times 45.5 \text{ ksi} \times 0.72 \times 16 \text{ in.}\right) = 2830 \text{ kip} \cdot \text{in.}$$

The computed capacity (2830 kip-in) is much smaller than the demand (4410 kip-in.). Because we are already estimating an excessive amount of required reinforcement (the amount of reinforcement we selected exceeds both ρ_b = 2.2% and ρ_{max} = 1.6%), we realize that the capacity cannot meet the demand unless we increase the dimensions of the section or use stronger concrete.

BARE ESSENTIALS

In general, beam reinforcement ratio should be between 0.2% and 2%.

EXERCISES

1. Repeat Example 16.1 for L = 20 ft.
2. Plot $\rho_{max} = \frac{3}{4}\,\rho_b$ vs. concrete cylinder strength (f'_c) for 3 ksi < f'_c < 6 ksi and f_y = 60 ksi.
3. Select dimensions and longitudinal reinforcement for the beam shown in Figure 16.6. Assume that the beam is required to resist a *factored* uniformly distributed load w_u = 2 kip-ft. Your choices should meet the limits defined in this section. Assume
   - The factored load given includes the self weight of the beam.
   - The strength reduction factor is $\Phi = 0.9$.
   - f_y = 60 ksi.
   - f'_c = 5000 psi.

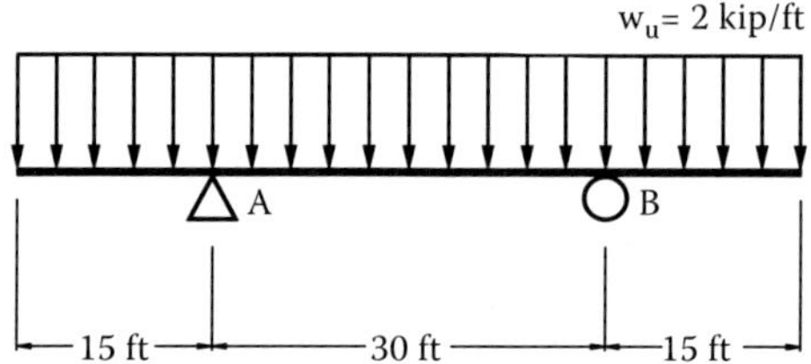

FIGURE 16.6 Beam with overhanging spans.

17 Beams with Compression Reinforcement

Compression reinforcement helps to

1. Reduce the effects of shrinkage and creep on beam deflections (Chapter 20)
2. Increase the useful limit of unit curvature

Compression reinforcement is not effective in increasing flexural strength of underreinforced sections. In this section, we study the effects of compression reinforcement on the useful limit of unit curvature and flexural strength.

We define the limit state by choosing the useful limit of compressive unit strain in the concrete to be 0.003. The distribution of strain is assumed to be linear (Figure 17.1). Given the linear strain distribution,

$$\varepsilon_s' = \frac{\varepsilon_{cu}}{k_u d} \cdot (k_u d - d') \tag{17.1}$$

Introducing a definition for convenience, $\gamma_c = d'/d$,

$$\varepsilon_s' = \frac{\varepsilon_{cu}}{k_u} \cdot (k_u - \gamma_c) \tag{17.2}$$

Rearranging terms,

$$k_u = \frac{\varepsilon_{cu}}{\varepsilon_{cu} - \varepsilon_s'} \gamma_c \tag{17.3}$$

The expression relating strains in the reinforcement in tension and the maximum compressive strain still holds:

$$k_u = \frac{\varepsilon_{cu}}{\varepsilon_{cu} + \varepsilon_{su}} \tag{17.4}$$

The expression for equilibrium of forces requires a new term describing the force carried by the reinforcement in compression (C_s):

$$T = C_c + C_s \tag{17.5}$$

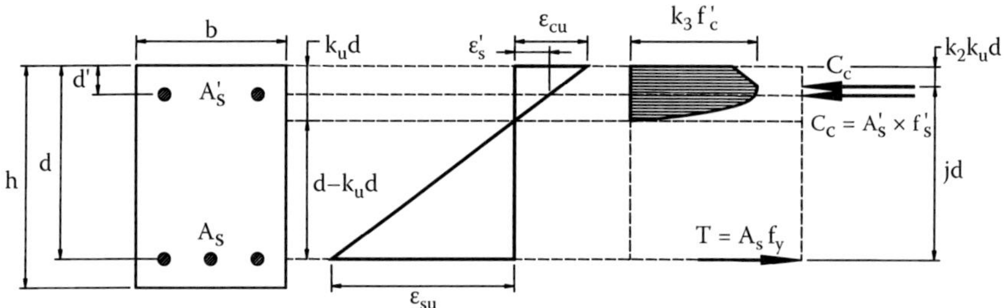

FIGURE 17.1 Strains, concrete stresses, and internal forces for a section with compression reinforcement.

where T is the tensile force, $T = A_s \cdot f_{sy}$ (if the steel in tension yields); C_c, the compressive force in the concrete, $= k_1 \cdot 0.85 \cdot f'_c \cdot (b \cdot k_u d - A'_s)$ (we subtract the area A'_s because it is occupied by steel); and C_s is the compressive force resisted by steel, $C_s = A'_s \cdot f'_s$.

Rewriting Equation 17.5,

$$A_s \cdot f_y = k_1 \cdot 0.85 \cdot f'_c \cdot b \cdot k_u d + A'_s \cdot \left(f'_s - k_1 \cdot 0.85 \cdot f'_c\right) \tag{17.6}$$

Rearranging terms and discarding the term $k_1 \times 0.85 \times f'_c$ (typically much smaller than f'_s),

$$k_u = \frac{\rho \cdot f_y - \rho' \cdot f'_s}{k_1 \cdot 0.85 \cdot f'_c} \tag{17.7}$$

where $\rho' = A'_s/(bd)$ is the compressive reinforcement ratio (note that we divide by d, *not* d′).

Equation 17.7 indicates that as the product $\rho' f'_s$ increases, the neutral axis depth decreases. Because limiting curvature is inversely proportional to neutral axis depth ($\phi = \varepsilon_c/(k_u d)$), we conclude that the limiting curvature increases with increasing amount of compression reinforcement as long as the depth to the neutral axis, $k_u d$, exceeds the depth of the compression reinforcement, d′. Displacements and signs warning us of potential failure also increase for beams with larger amounts of compression reinforcement. From Equation 17.7 it can also be inferred that addition of compression reinforcement will permit the use of more tensile reinforcement without exceeding the limits discussed in Chapter 16.

The unit stress in the compression reinforcement must decrease as ρ′ increases and k_u decreases. To estimate the unit stress in the compression reinforcement (f'_s) we combine

$$k_u = \frac{\rho \cdot f_y - \rho' \cdot f'_s}{k_1 \cdot 0.85 \cdot f'_c}$$

and

$$k_u = \frac{\varepsilon_{cu}}{\varepsilon_{cu} - \varepsilon_s'} \gamma_c$$

and obtain f_s' as a function of ε_s':

$$f_s' = \frac{1}{\rho'} \left(\rho \cdot f_y - \frac{\varepsilon_{cu}}{\varepsilon_{cu} - \varepsilon_s'} \gamma_c \cdot k_1 \cdot 0.85 \cdot f_c' \right) \tag{17.8}$$

As we have done previously, we superimpose the curve described by Equation 17.8 and $f_s' = E_s\, \varepsilon_s' \leq f_y$. The intersection gives us an estimate of f_s' and ε_s' that satisfies all the expressions in our formulation. The student is encouraged to explore the formulation presented to see that as we increase ρ', f_s' and ε_s' decrease, and that if we limit ρ to 2% (to avoid congestion), it is rather difficult to reach yielding in compression reinforcement.

Having estimates of f_s' and ε_s' allows us to compute k_u and check our assumption about the stress in the tension reinforcement ($f_s = f_y$, which is the case if k_u<3/5 and f_y = 60 ksi). It allows also us to estimate moment capacity as

$$M_n = C_c \cdot \left(1 - \tfrac{1}{2} k_1 \cdot k_u\right) \cdot d + C_s \cdot (d - d') \tag{17.9}$$

In the above expression, moment is determined about the tension reinforcement. The terms defining the moment are those representing the compressive forces in the concrete and in the compression reinforcement. It is important to note that the moment computed is close to $0.9A_s f_y d$ unless the tensile reinforcement ratio is excessive.

EXAMPLE 17.1

Compute nominal moment capacity and limiting curvature for the section shown in Figure 17.2 and the following parameters:

ρ = 1.5%

f_c' = 5000 psi

f_y = 60 ksi

γ_c = 0.1

ρ' = 0, 0.5%, and 1%

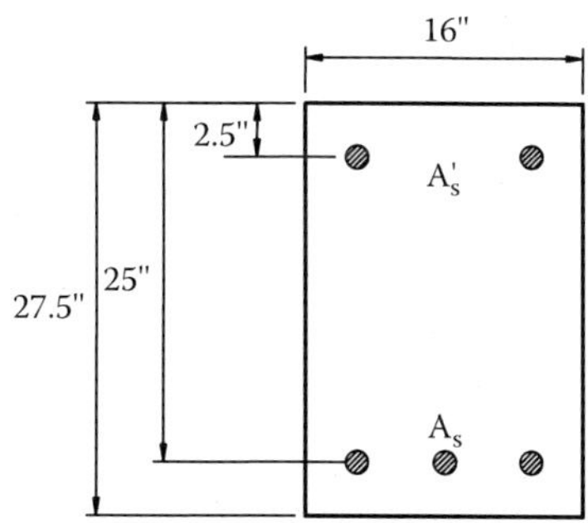

FIGURE 17.2 Example 17.1.

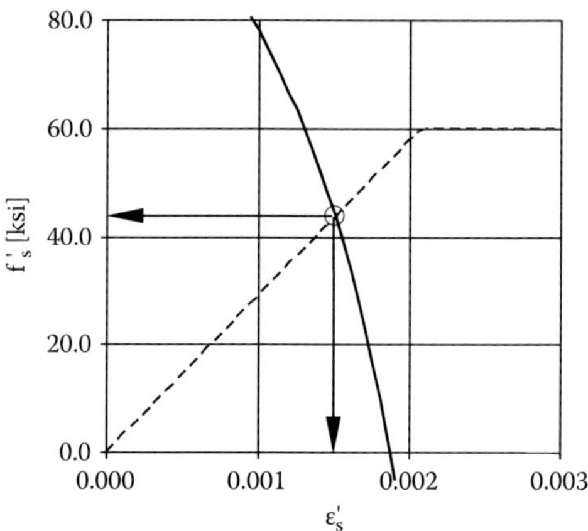

FIGURE 17.3 $\rho' = 0.5\%$.

SOLUTION

For the case with $\rho' = 0$:

$$k_u = \frac{\rho \cdot f_y}{k_1 \cdot 0.85 \cdot f_c'} = \frac{0.015 \cdot 60 \text{ ksi}}{0.8 \cdot 0.85 \cdot 5 \text{ ksi}} = 0.26 \; (<3/5 \Rightarrow f_s = f_y)$$

$$\phi_u = \frac{\varepsilon_{cu}}{k_u \cdot d} = \frac{0.003}{0.26 \cdot 25 \text{ in.}} \boxed{= 4.6 \times 10^{-4} \text{ 1/in.}}$$

$$j = 1 - \frac{1}{2} \cdot k_1 \cdot k_u = 1 - \frac{1}{2} \cdot 0.8 \cdot 0.26 = 0.9$$

$$M_n = \rho \cdot b \cdot d \cdot f_y \cdot jd = 1.5\% \cdot 16 \text{ in.} \cdot 25 \text{ in.} \cdot 60 \text{ ksi} \cdot 0.9 \cdot 25 \text{ in.} \boxed{= 8100 \text{ kip} \cdot \text{in.}}$$

For the case with $\rho' = 0.5\%$:

$$\frac{\rho \cdot f_y - \rho' \cdot f_s'}{k_1 \cdot 0.85 \cdot f_c'} = \frac{\varepsilon_{cu} \cdot}{\varepsilon_{cu} - \varepsilon_s'} \gamma_c \text{ leads to}$$

$$f_s' = \frac{1}{0.005}\left(0.015 \cdot 60 \text{ ksi} - \frac{0.003}{0.003 - \varepsilon_s'} 0.1 \cdot 0.8 \cdot 0.85 \cdot 5 \text{ ksi}\right)$$

We compare this equation with the unit stress–unit strain relationship assumed for steel and obtain $f_s' = 44$ ksi and $\varepsilon_s' = 0.0015$ (Figure 17.3). With this estimate we compute

$$k_u = \frac{\varepsilon_{cu} \cdot}{\varepsilon_{cu} - \varepsilon_s'} \gamma_c = \frac{0.003}{0.003 - 0.0015} 0.1 = 0.2 \; (<3/5)$$

$$\phi_u = \frac{\varepsilon_{cu}}{k_u \cdot d} = \frac{0.003}{0.2 \cdot 25 \text{ in.}} \boxed{= 6 \times 10^{-4} \text{ 1/in.}}$$

$$M_n = k_1 \cdot 0.85 \cdot f'_c \cdot b \cdot k_u d^2 \cdot \left(1 - \tfrac{1}{2} k_1 \cdot k_u\right) + A'_s \cdot f'_s \cdot (d - d')$$

$$M_n = 0.8 \cdot 0.85 \cdot 5 \text{ ksi} \cdot 16 \text{ in.} \cdot 0.2 \cdot (25 \text{ in.})^2 \cdot \left(1 - \tfrac{1}{2} 0.8 \cdot 0.2\right)$$

$$+ 0.5\% \cdot 16 \text{ in.} \cdot 25 \text{ in.} \cdot 44 \text{ ksi} \cdot (25 - 2.5) \text{ in.}$$

$$\approx 6250 \text{ kip} \times \text{in.} + 1980 \text{ kip} \times \text{in.} \approx \boxed{8200 \text{ kip} \times \text{in.}}$$

For the case with $\rho' = 1\%$:

$$\frac{\rho \cdot f_y - \rho' \cdot f'_s}{k_1 \cdot 0.85 \cdot f'_c} = \frac{\varepsilon_{cu} \cdot}{\varepsilon_{cu} - \varepsilon'_s} \gamma_c \quad \text{leads to}$$

$$f'_s = \frac{1}{0.01}\left(0.015 \cdot 60 \text{ ksi} - \frac{0.003}{0.003 - \varepsilon'_s} 0.1 \cdot 0.8 \cdot 0.85 \cdot 5 \text{ ksi}\right)$$

We compare this equation with the unit stress–unit strain relationship assumed for steel and obtain $f'_s = 34$ ksi and $\varepsilon'_s = 0.0012$ (Figure 17.4). Now

$$k_u = \frac{\varepsilon_{cu} \cdot}{\varepsilon_{cu} - \varepsilon'_s} \gamma_c = \frac{0.003}{0.003 - 0.0012} 0.1 = 0.17 \ (<3/5)$$

$$\phi_u = \frac{\varepsilon_{cu}}{k_u \cdot d} = \frac{0.003}{0.17 \cdot 25 \text{ in.}} \boxed{= 7 \times 10^{-4} \text{ 1/in.}}$$

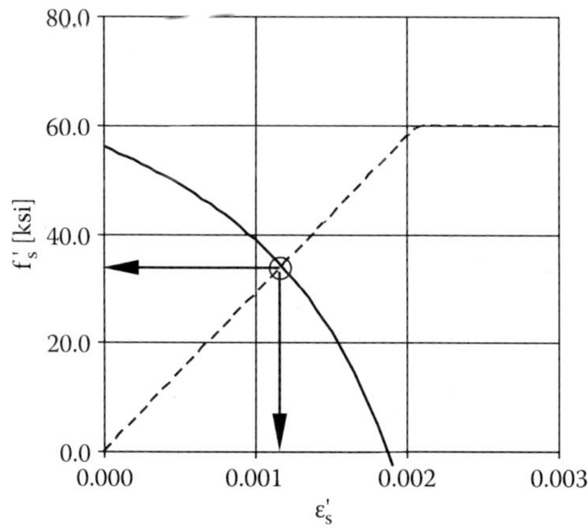

FIGURE 17.4 $\rho' = 1.0\%$.

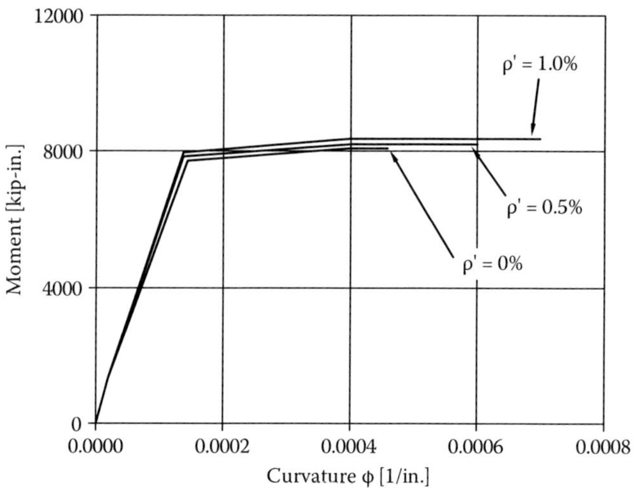

FIGURE 17.5 Moment–curvature diagrams.

$$M_n = k_1 \cdot 0.85 \cdot f_c' \cdot b \cdot k_u d^2 \cdot \left(1 - \tfrac{1}{2} k_1 \cdot k_u\right) + A_s' \cdot f_s' \cdot (d - d')$$

$$M_n = 0.8 \cdot 0.85 \cdot 5 \text{ ksi} \cdot 16 \text{ in.} \cdot 0.17 \cdot (25 \text{ in.})^2 \cdot \left(1 - \tfrac{1}{2} 0.8 \cdot 0.17\right)$$

$$+1\% \cdot 16 \text{ in.} \cdot 25 \text{ in.} \cdot 34 \text{ ksi} \cdot (25 - 2.5) \text{ in.}$$

$$M_n = 5380 \text{ kip} \cdot \text{in.} + 3060 \text{ kip} \cdot \text{in.} \boxed{\approx 8400 \text{ kip} \cdot \text{in.}}$$

Our results and other moment–curvature pairs are plotted in Figure 17.5. The strength increases associated with the addition of compression reinforcement are virtually negligible. Nevertheless, in the case analyzed, adding 0.5% compression reinforcement increased the calculated limiting unit curvature by 30%. Adding another 0.5% compression reinforcement led to an additional 20% increase in the calculated limiting curvature.

BARE ESSENTIALS

- Compression reinforcement is effective in reducing time-dependent deflection in beams.
- Compression reinforcement, obtained by hoops, can be used to increase the deformation capacity in flexure but it is not effective in increasing flexural strength.

EXERCISE

Compute the moment capacity and the limiting curvature of the section described in the second example of Chapter 14 assuming that five Grade 60 #9 bars are added at 3 in. from the top of the beam. Compare your results with the results from Chapter 14. In a single sentence state what can be concluded from this comparison.

18 Beams with Flanges

Cast-in-place reinforced concrete structures are typically monolithic. Casting is accomplished through a series of individual lifts, but the entire structure is one continuous element as long as the reinforcing details effect continuity. A slab cast on a beam (Figure 18.1) becomes the top flange of the beam. Such a beam is said to have a T-section.*

Chapter 18 focuses on the proportioning of beams with T-sections. We are concerned with the determinations of the moment and unit curvature capacities of the section. Before we tackle the subject, there are four issues that we should note:

1. The flexural response of a T-section subjected to negative bending moment (top flange in tension) is essentially the same as that of a rectangular section except for the moment and curvature at cracking. The moments and curvatures at yield and at the limiting condition may be determined as described in Chapter 14. It is important to note that the T-beam flange can accommodate tensile reinforcement conveniently.
2. In T-sections subjected to positive bending moment (top flange in compression), if the depth to the neutral axis, at yield and at the limit, calculated using the expressions for a rectangular section is within the flange thickness, t_f, flexural response at yield and at the limit can be determined using the expressions for rectangular sections.
3. If the designer elects to ignore the effect of the top flange on the strength of the compression zone in determining the required amount of tensile reinforcement, it is highly unlikely that any existing code authority will fault the designer.
4. In cases where the beam spacing is large, the designer needs to define the effective width of the top flange (Figure 18.2). A pragmatic rule is to assume that the flange on each side is equal to the depth of the beam below the slab. The thickness of the slab also influences the flange width. Current building codes tend to limit the flange, on each side, to less than three to eight times the slab thickness. The range of opinion reflects the variability of results of analysis and judgment. For light to moderate tensile reinforcement, decisions about the flange width make very little difference on proportioning of the section, as implied by the freedom given to the designer for ignoring the flange. Of course, the flange width may not be assumed to extend more than

* If the beam is located at the edge of the slab, it is said to have an L-section. If the beam has two flanges at its top and bottom ends, it is said to have an I-section.

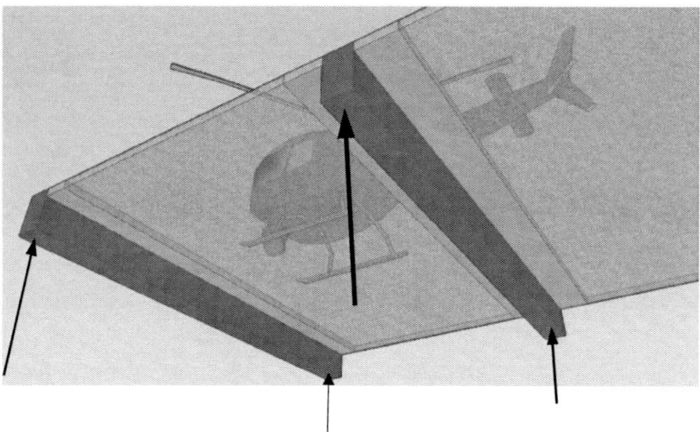

FIGURE 18.1 Beams supporting a slab.

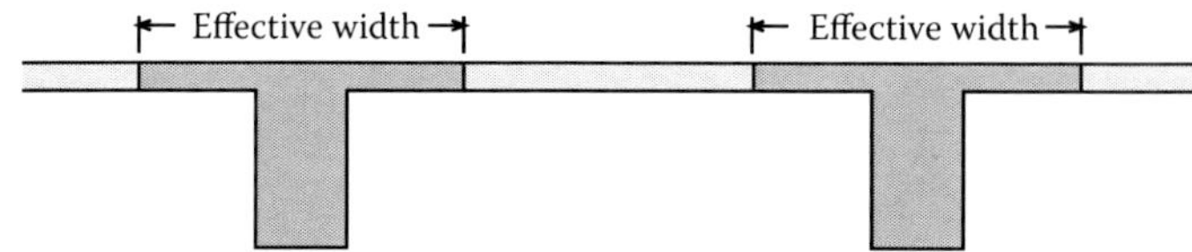

FIGURE 18.2 Effective width.

half the clear distance to the next parallel beam. In all cases, the choice of the flange width dimension must not exceed the maximum allowed by the applicable code.

18.1 A T-SECTION SUBJECTED TO POSITIVE MOMENT

To determine the moment and curvature capacity, we assume that the strain distribution is linear over the section, and we project this assumption to hold for the overhanging portions of the flange.

If the neutral axis falls within the flange ($k_u d < t_f$), the expressions derived for rectangular sections apply except that the width of the area of concrete in compression is not b, but b_f (Figure 18.3). Therefore, we redefine ρ:

$$\rho_f = \frac{A_s}{b_f d} \tag{18.1}$$

We determine the relative depth to the neutral axis from equilibrium of normal forces on the section as we did for a rectangular section:

$$k_u = \frac{\rho_f \cdot f_y}{k_1 f_c''} \tag{18.2}$$

With this definition of k_u, the expressions derived in Chapter 14 for the relative internal moment arm (j) and moment capacity of rectangular beams remain applicable.

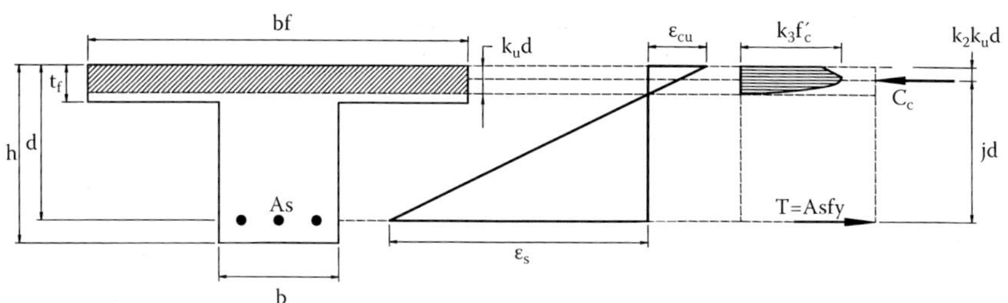

FIGURE 18.3 Cross section, strain distribution, concrete stresses, and internal forces.

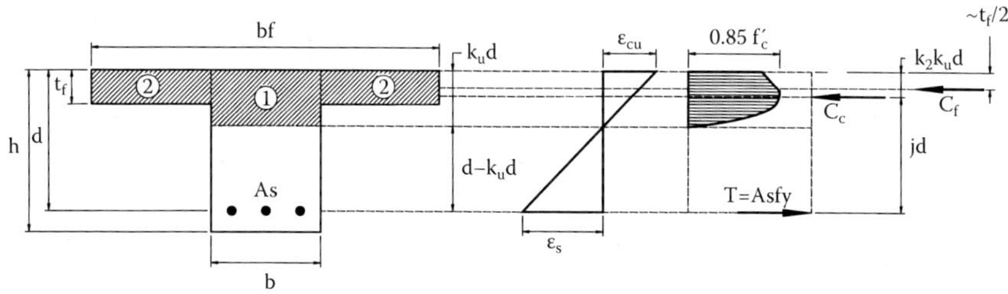

FIGURE 18.4 Cross section, strain distribution, concrete stresses, and internal forces ($k_1 k_u d > t_f$).

The designer should realize that ρ_f is a fraction of ρ, and it is a poor indicator of whether the reinforcement will fit within the web.

If the calculated ratio k_u exceeds the ratio t_f/d, to be consistent with assumptions made, then we divide the area in compression into two regions (Figure 18.4). The resultant of stresses in region 1 (within the web) is determined as the product of the mean compressive stress in the concrete, $k_1 0.85 f'_c$, the web thickness, b, and the depth to the neutral axis, $k_u d$.

$$C_c = k_1 \cdot 0.85 f'_c \cdot b \cdot k_u d \tag{18.3}$$

If we consider the assumptions made in determining the flange width and the variation of stress to be accurate, determining the resultant of stresses in region 2 requires detailed work because it depends on the ratio of flange thickness to neutral axis depth. Because the error introduced is small, and because the width b_f is not exact, we simply assume the force on region 2 to be

$$C_f = 0.85 f'_c \cdot (b_f - b) \cdot t_f \tag{18.4}$$

In essence, our assumption implies that the two three-dimensional objects shown in Figure 18.5 have the same volume. They do not. The "wings" of the lower object are too large. The centroids of these objects do not coincide either. We ignore these

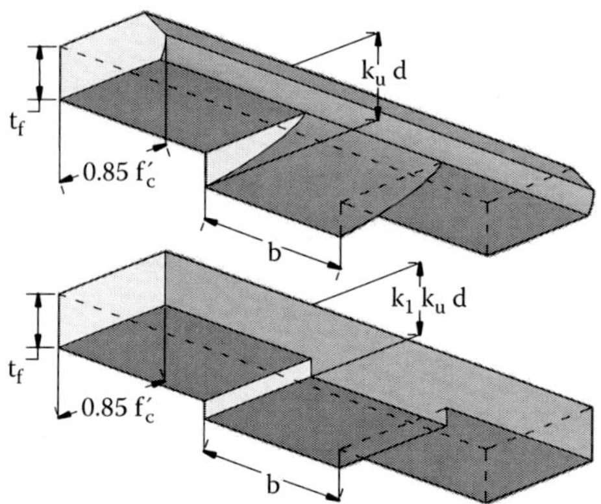

FIGURE 18.5 Stress distributions.

discrepancies to write an expression describing equilibrium of axial forces similar to the expression we wrote for beams with compression reinforcement:

$$C_c + C_f = T \tag{18.5}$$

$$k_1 \cdot 0.85f_c' \cdot b \cdot k_u d + 0.85f_c' \cdot (b_f - b) \cdot t_f = A_s f_y \tag{18.6}$$

Solving for k_u and simplifying,

$$k_u = \frac{\rho \cdot f_y}{k_1 \cdot 0.85f_c'} - \frac{1}{k_1}\frac{t_f}{d}\left(\frac{b_f}{b} - 1\right) \tag{18.7}$$

The term

$$\frac{1}{k_1}\frac{t_f}{d}\left(\frac{b_f}{b} - 1\right)$$

represents a shift of the neutral axis caused by the effect of the slab. If this shift is large enough to make the product $k_1 k_u d$ smaller than t_f, then we apply again the expressions derived for rectangular beams (as modified above).

For $\rho < 2\%$: $f_y = 60$ ksi, $f_c' > 4$ ksi. The term

$$\frac{\rho \cdot f_y}{k_1 \cdot 0.85f_c'}$$

is smaller than ~2/5. We replace this value (2/5) in the expression we derived for k_u and conclude that, within the defined ranges, if t_f exceeds approximately

$$\frac{2}{5} \cdot k_1 \frac{b}{b_f} d,$$

then the depth of the equivalent compression block ($k_1 k_u d$) is *smaller* than the thickness of the flange (t_f) and we can treat the section as if it were rectangular.

The function

$$\frac{2}{5}\cdot k_1\frac{b}{b_f}$$

is plotted vs.

$$\frac{b}{b_f}$$

for

$$\frac{b}{b_f}<\frac{1}{3}$$

and $k_1 = 0.85$ in Figure 18.6. From this plot we conclude that, within the ranges we have defined, T-beams with t_f/d larger than 1/8 can be treated as rectangular beams (using a revised reinforcement ratio):

$$\rho_f=\frac{A_s}{b_f\cdot d}$$

We need to examine other cases more closely. If the depth of the equivalent stress block falls outside the flange, moment capacity is computed as

$$M_n=C_c\cdot\left(1-\tfrac{1}{2}k_1\cdot k_u\right)\cdot d+C_f\cdot\left(d-\tfrac{1}{2}t_f\right) \tag{18.8}$$

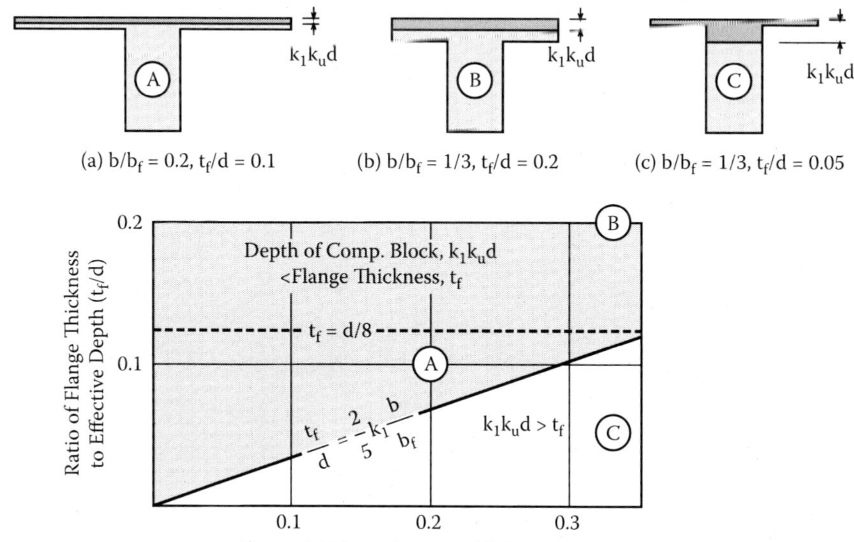

FIGURE 18.6 Domain in which the depth of the compression block is smaller than the flange thickness.

The moment, M_n, can be estimated closely as $M_n = A_s \cdot f_g \cdot 0.9d$ unless the section is very heavily reinforced.

EXAMPLE 18.1

Compute the moment capacity of the section shown in Figure 18.7 assuming $f'_c = 5000$ psi.

SOLUTION

If we assume that the section can be treated as if it were rectangular, we compute

$$\rho = \frac{A_s}{b_f \cdot d} = \frac{3 \text{ in.}^2}{48 \text{ in.} \cdot 21 \text{ in.}} = 0.3\%$$

and

$$k_u = \frac{\rho \cdot f_y}{k_1 \cdot 0.85\, f'_c} = \frac{0.003 \cdot 60 \text{ ksi}}{0.8 \cdot 0.85 \cdot 5 \text{ ksi}} = 0.053 < 0.6 \text{ (and, therefore, } f_s = f_y)$$

The depth of the equivalent compression block would be $k_1 k_u\, d = (0.8 \times 0.053) \times 21$ in. $= 0.042 \times 21$ in. $= 0.9$ in. This is (much) smaller than the thickness of the flange $t_f = 6$ in. We should have expected this result because $t_f/d = 0.29 > 1/8$ and

$$\frac{b}{b_f} < \frac{1}{3}$$

The internal lever arm is

$$j = 1 - k_1 k_u/2 = 1 - 0.02 = 0.98$$

The nominal moment capacity is

$$M_n = A_s f_y jd = 3 \text{ in.}^2 \cdot 60 \text{ ksi} \cdot 0.98 \cdot 21 \text{ in.} = 3700 \text{ kip} \cdot \text{in.}$$

Note that this result is again close (the error is 8%) to $A_s f_y\, 0.9d = 3 \text{ in.}^2 \cdot 60 \text{ ksi} \cdot 0.9 \cdot 21 \text{ in.} = 3400$ kip in.

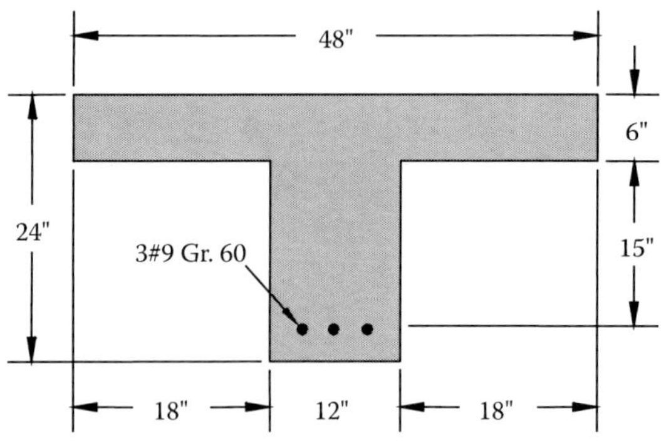

FIGURE 18.7 Example 18.1.

BARE ESSENTIALS

- If $t_f > d/8$ and $b/b_f < 1/3$, we can usually treat beams with T-shaped sections as beams with rectangular sections with a reinforcement ratio

$$\rho_f = \frac{A_s}{b_f \cdot d}$$

- In design, it is not indefensible or unsafe to ignore the effect of the flange and treat the section as a rectangular section having a width equal to that of its web if the flange is in compression.

Remember: The limiting curvature decreases with increases in reinforcement ratio.

EXERCISE

Compute the moment capacity and limiting curvature of the beam section shown below. The section is subjected to positive moment and $f'_c = 3000$ psi. Sketch the 3D stress distribution in the concrete (as in Figure 18.5).

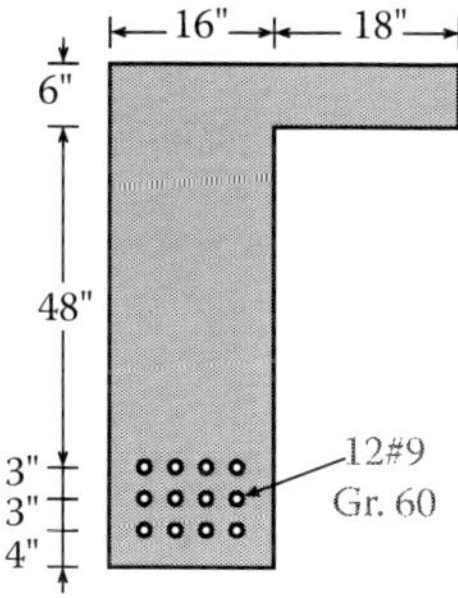

19 Deflection under Short-Time Loading

Deflections of reinforced concrete elements subjected to flexure create design challenges mainly in relation to six issues:

1. Damage to partition walls on which they may come to rest
2. Restraint of door mechanisms in contact with flexural elements
3. Damage to piping attached near mid-span of flexural elements
4. Ponding of water in roof slabs
5. Unevenness
6. Flexibility of floor slabs leading to perceptible vibrations

Deflection is primarily a serviceability and seldom a safety concern. For a given static loading, the main factors controlling service load deflection are the magnitude of reinforcement stress at service load and the effective depth of the section. To keep deflection low for a given bending moment demand and span length, the basic thinking process is simple. The product of the cross-sectional area of the tensile reinforcement and the effective depth ($A_s \cdot d$) should be kept as high as possible within economic limits and without surpassing the limits related to constructability and the maximum reinforcement ratio (Chapter 16).

From the viewpoint of efficient and proper design, it is essential that the deflection problem be approached as a problem in geometry. Compact expressions relating deflection to load directly in terms of Young's modulus, moment of inertia, and span (such as $\Delta = 5wL^4/(384EI)$) for a uniformly loaded and simply supported prismatic beam may save time but do not help with understanding the physical process. In this section we develop a simple procedure to determine the short-time deflection of reinforced concrete elements with transverse loads. Short-time deflection refers to deflection that is not affected by time-dependent volume changes in concrete. Strictly, we are thinking of a load that is applied in a matter of minutes but not milliseconds or days. Our goal will be to develop procedures to estimate and, more importantly, understand the deflection response of slender reinforced concrete beams.

To begin with, we need to emphasize that in structural analysis related to deflection, the object is not to predict deflection. Rather, the object is to control deflection or to keep it within bounds to avoid the serviceability problems listed above.

To determine the proper depth of the beam should be very simple. All one has to do is to look at existing beams and develop a sensitivity to their span-to-depth ratios. Naturally, the successful span-to-depth ratio will be different for different boundary conditions and for different load magnitudes and distributions. However, for a large number of cases the student will observe that depths of reinforced concrete beams range from 1/10 to 1/14 of the clear span unless loads are light as in the case of joists. Even though it may appear to be an argument for "knownothingism," the budding designer would be wise to bear in mind that a calculated deflection is nothing more than a guess. It is better to put one's faith in proportions that have led to success in the past rather than believe that a calculated beam deflection is exact.

19.1 DEFLECTION OF A BEAM SUBJECTED TO BENDING MOMENT

Determination of the deflection of a beam is best handled in relation to its strain geometry. To develop the correct perspective for treating the deflection problem in geometry, we start with a very simple case: a prismatic and homogeneous beam segment subjected to constant moment, as shown in Figure 19.1a. We assume that the beam is weightless (no vertical reactions needed and no change in moment along the span), and that it is not stressed beyond its linear range of response.

To help visualize the deflection phenomenon, we break the segment into ten elements of equal length (Figure 19.1b).

First, we consider element 1 (to the left of mid-span in Figure 19.1c). We start from mid-span because we know, from the principle of symmetry, that the slope of the beam is zero at mid-span. The applied moment results in constant unit curvature, ϕ, over the length of the element. The corresponding angle change is the product of the unit curvature ϕ and the length of the element, L/10.

$$\theta_1 = \phi \cdot \frac{L}{10} \tag{19.1}$$

where ϕ is the unit curvature in response to applied moment M and L is the length of the beam.

The slope at the left end of element 1 is the slope at the right end (which we know to be zero) plus the angle change defined in Equation 19.1. We rotate the next element on the left (element 2) so that its right boundary coincides with the left boundary of the first element. The increase in angle change from the right to the left boundary of element 2 is the same as that of element 1. The angle change at the left boundary of element 2 is the sum of the angle change in element 1 plus that in element 2.

$$\theta_2 = \phi \cdot \frac{L}{10} + \phi \cdot \frac{L}{10} = \phi \cdot \frac{L}{5} \tag{19.2}$$

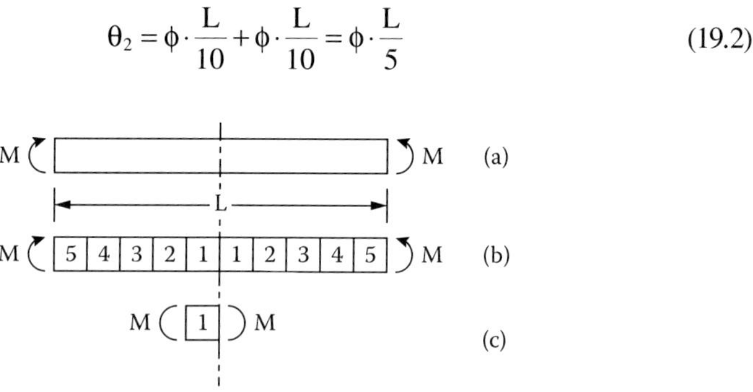

FIGURE 19.1 Prismatic beam segment subjected to constant moment.

Because the unit curvature in all elements is the same, it follows that the slopes at the ends of the beam can be obtained by summing the angle changes in the elements contributing to the slope at the end.

$$\theta_{left} = \theta_{right} = 5 \cdot \phi \cdot \frac{L}{10} = \frac{1}{2} \phi \cdot L \tag{19.3}$$

This derivation considering changes in the geometries of the discrete elements is, in essence, the same as the definition in the moment-area method that specifies, in simple terms, that the change in rotation from a point in a beam to another point is the integral of the unit curvature variation between the two points.

Using the moment-area method, we note that the rotation (or slope) at mid-span is zero and conclude that the rotation at the support would be simply the area of the unit curvature diagram over one half of the beam considered:

$$\theta_{support} = \frac{1}{2} \phi \cdot L \tag{19.4}$$

To determine the deflection at mid-span, we refer to Figure 19.2. We obtain the total deflection in a step-by-step procedure by adding the tangential offsets at the left support. The segments are numbered sequentially from center to the left support.

Deflection increments related to individual segments are

$$\Delta_1 = \frac{\phi \cdot L}{10} 4.5 \cdot \frac{L}{10}$$

$$\Delta_2 = \frac{\phi \cdot L}{10} 3.5 \cdot \frac{L}{10}$$

$$\Delta_3 = \frac{\phi \cdot L}{10} 2.5 \cdot \frac{L}{10}$$

$$\Delta_4 = \frac{\phi \cdot L}{10} 1.5 \cdot \frac{L}{10}$$

$$\Delta_5 = \frac{\phi \cdot L}{10} 0.5 \cdot \frac{L}{10}$$

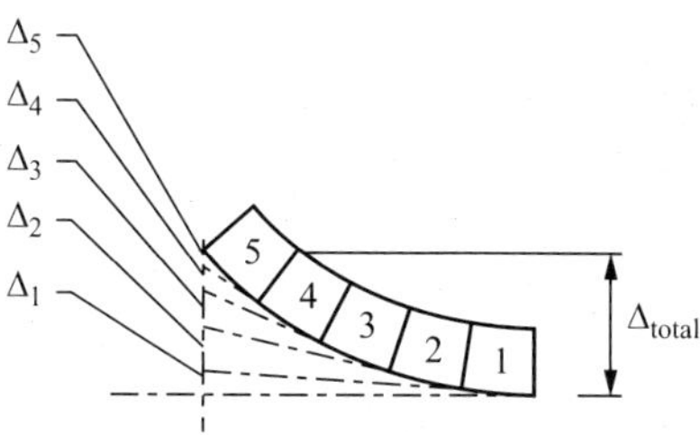

FIGURE 19.2 Deflection at mid-span.

The total deflection is

$$\Delta_{total} = \Delta_1 + \Delta_2 + \Delta_3 + \Delta_4 + \Delta_5 = \frac{\phi \cdot L^2}{100} \cdot (4.5 + 3.5 + 2.5 + 1.5 + 0.5).$$

$$\Delta_{total} = \frac{1}{8} \cdot \phi \cdot L^2 \tag{19.5}$$

We can arrive at the same result directly by using the moment-area method. For a beam with zero slope at mid-span, the deflection at mid-span is the deviation at either support of a tangent at mid-span. The cited deviation is determined as the first moment of the area of the unit curvature diagram between the mid-span and either support. In this simple case (with constant unit curvature along the entire beam):

Deflection at mid-span = Unit curvature × Tributary span × Distance from support to centroid of the unit curvature diagram

$$\Delta_{total} = \phi \cdot \frac{L}{2} \cdot \frac{L}{4} \tag{19.6}$$

$$\Delta_{total} = \frac{1}{8} \cdot \phi \cdot L^2 \tag{19.7}$$

The moment-area method is very simple and straightforward. It becomes a powerful and reliable tool as long as the engineer does not lose his or her understanding of the geometry leading to it.

It is to be understood that the method is to be used in relation to small deformations.*

19.2 DEFLECTION OF AN UNCRACKED REINFORCED CONCRETE BEAM WITH CONCENTRATED LOADS

Having discussed the fundamental aspects of the moment-area method and having emphasized that the deflection must always be considered in terms of strain and beam geometry, it is time to investigate beam deflections to check how large or small they are likely to be.

We start by considering an uncracked reinforced concrete beam. In practice an uncracked beam occurs very rarely, but obtaining an estimate of its deflection compared with its span is instructive.

Figure 19.3a shows a simply supported beam spanning over a distance L with equal loads at third points of its span. The beam reactions are each (ignoring self weight)

$$R_A = \frac{P}{2} \tag{19.8}$$

* Not exceeding 3% of the span.

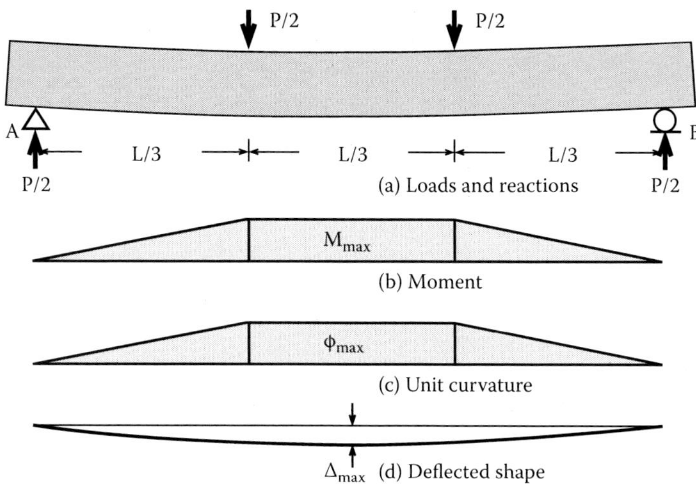

FIGURE 19.3 Simply supported beam with equal loads at third-points of its span.

where R_A is the reaction at the left support in response to applied load P and P is the total applied load, resulting in a maximum moment in terms of P and L of

$$M_{max} = \frac{P}{2} \cdot \frac{L}{3} = \frac{P \cdot L}{6} \tag{19.9}$$

where M_{max} is the moment acting over the middle third of the beam span.

To treat the problem in terms of strain geometry, we determine the maximum stress corresponding to the maximum moment

$$f_c = \frac{M \cdot \frac{h}{2}}{I} = \frac{PL}{6} \cdot \frac{h}{2} \cdot \frac{1}{\frac{1}{12} b \cdot h^3} = \frac{P \cdot L}{b \cdot h^2} \tag{19.10}$$

where h is the total height of the section, b is the width of the rectangular section, and I is the moment of inertia of the rectangular section.

From the maximum stress, we obtain the maximum strain

$$\varepsilon_c = \frac{f_c}{E_c} = \frac{P \cdot L}{E_c \cdot b \cdot h^2} \tag{19.11}$$

where ε_c is the strain at the extreme fiber, f_c is the stress at the extreme fiber, and E_c is Young's modulus for beam material.

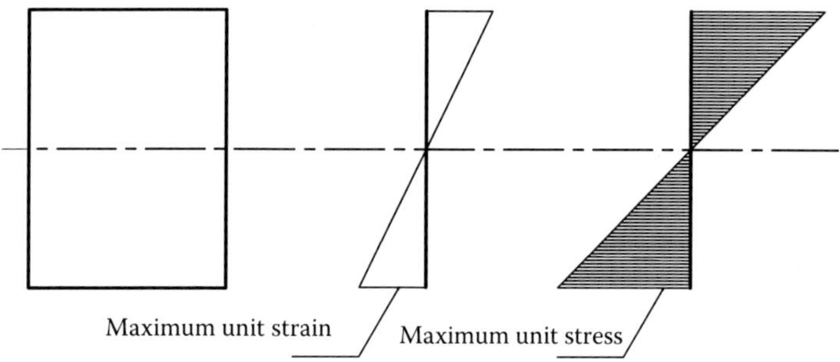

FIGURE 19.4 Strain and stress distribution of homogeneous beam.

For a linear distribution of strain over the height of the homogeneous beam, (Figure 19.4) the maximum unit curvature is

$$\phi_{max} = \frac{\varepsilon_c}{\frac{h}{2}} = 2\frac{\varepsilon_c}{h} \tag{19.12}$$

where ϕ_{max} is the unit curvature corresponding to M_{max}.

Using the moment-area method, we calculate the deviation at one support of a tangent at mid-span to obtain the deflection at mid-span:

$$\Delta_{max} = \left(\phi_{max} \cdot \frac{L}{3} \cdot \frac{1}{2}\right) \cdot \left(\frac{2}{3} \cdot \frac{L}{3}\right) + \left(\phi_{max} \cdot \frac{L}{6}\right) \cdot \left(\frac{L}{3} + \frac{L}{12}\right) \tag{19.13}$$

$$\Delta_{max} = \frac{23}{216} \cdot \phi_{max} \cdot L^2 = \frac{23}{216} \cdot \left(2 \cdot \frac{\varepsilon_c}{h}\right) \cdot L^2 \tag{19.14}$$

The expression refers to a homogeneous beam responding to load within its linear range of response (an uncracked concrete beam). Next, we deal with a cracked reinforced concrete section.

19.3 DEFLECTION OF A CRACKED REINFORCED CONCRETE BEAM WITH CONCENTRATED LOADS

The strain geometry of the cracked section is shown in Figure 19.5. In Chapter 13, the value of k was derived to be (in terms of the section properties for a rectangular section reinforced in tension only)

$$k = \sqrt{(\rho n)^2 + 2\rho n} - \rho n$$

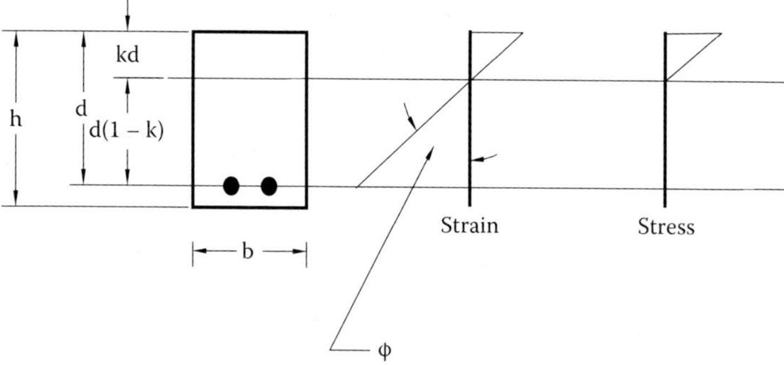

FIGURE 19.5 (a) Cross section. (b) Unit strain. (c) Unit stress in concrete.

We assume that the section is reinforced in tension only, and we ignore the tensile strength of concrete. From the strain geometry we obtain the relationship

$$\frac{\varepsilon_c}{k \cdot d} = \frac{\varepsilon_s}{d \cdot (1-k)} \tag{19.15}$$

where d is the depth from the extreme fiber in compression to the geometric centroid of tensile reinforcement (effective depth of section); n is the modular ratio, the ratio of Young's modulus for steel to that of concrete; k is the ratio of the depth from the extreme fiber in compression to the neutral axis to depth d; ε_c is the concrete strain at the extreme fiber in compression; and ε_s is the strain at the centroid of tensile reinforcement.

With k expressed in terms of the properties of the section, we define the maximum unit curvature for a rectangular concrete section reinforced in tension and responding in its linear range of response as

$$\phi_{max} = \frac{\varepsilon_s}{d \cdot (1-k)} \tag{19.16}$$

Substituting the curvature in the expression for deflection, deflection at mid-span is

$$\Delta_{max} = \frac{23}{216} \phi_{max} \cdot L^2 = \frac{23}{216} \cdot \left(\frac{\varepsilon_s}{d \cdot (1-k)} \right) \cdot L^2 \tag{19.17}$$

EXAMPLE 19.1

To develop a sense of proportion, we evaluate the expression above using representative dimensions. We assume the span L to be 21 ft (Figure 19.6a). Having looked at other simply supported beams, we make the depth to be approximately

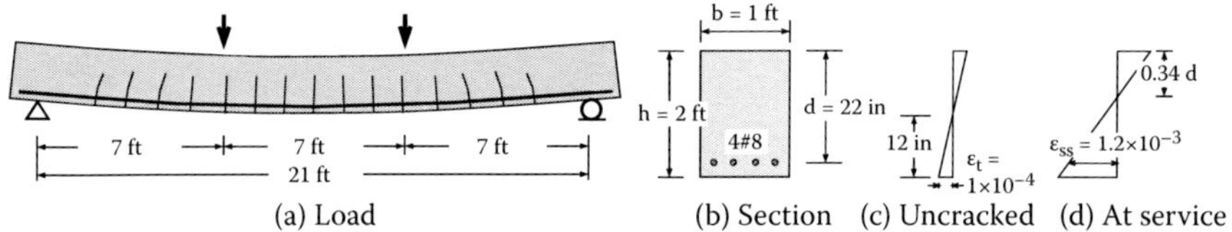

FIGURE 19.6 Moment distribution.

one-tenth of the span, or 2 ft (Figure 19.6b). Again, looking at sections we have seen in textbooks and around us, we set the width of the section to be 1 ft.

First, we consider the beam in its uncracked state. To simplify the solution, we ignore the influence of the tensile reinforcement on stiffness. For the section to remain uncracked, the tensile stress of the concrete should not be exceeded. We consider a tolerable tensile stress of 400 psi. We assume the Young's modulus for the concrete to be E_c = 4000 ksi. With that stress in the extreme fiber, the corresponding strain is (Figure 19.6c)

$$\varepsilon_{cr} = \frac{f_r}{E_c} = 1 \times 10^{-4}$$

The maximum unit curvature is

$$\phi_{max} = \frac{\varepsilon_{cr}}{\frac{h}{2}} = 8.3 \times 10^{-6} \cdot \frac{1}{in.}$$

The deflection at midspan is

$$\Delta_{max} = \frac{23}{216} \cdot \phi_{max} \cdot L^2 = 0.06 \text{ in.}$$

In terms of the span, the deflection amounts to

$$\frac{\Delta_{max}}{L} = 0.02\%$$

The ratio of the span to the calculated deflection is

$$\frac{L}{\Delta_{max}} = 4.5 \times 10^3$$

These ratios tell us that the deflection is negligibly small. The conclusion does not surprise us. What we have calculated is an approximation to the deflection at first cracking of a reinforced concrete girder spanning 20 ft. It should be very small. We should also remember that the computed deflection is qualified by assumptions we have made in arriving at it. We ignored the self weight and the presence of tensile reinforcement. We ignored any residual stresses (related to temperature or shrinkage) that may exist in the concrete. In fact, the deflection we have computed is not of much practical significance, but it is useful for us to understand the effect of tensile strength of concrete on deflections, a question we consider next.

It is interesting to note that we could have arrived at the maximum deflection in a simpler way, as follows.

It is known from experiments that, for normal weight aggregate concrete, the cracking strain ranges from 0.0001 to 0.0002. We take the lower bound to determine the unit curvature:

$$F_{max} = 0.0001/(24 \cdot \text{in.}/2) = 8 \times 10^{6}\ 1/\text{in.}$$

Using the expression for maximum deflection stated in terms of F_{max},

$$\Delta_{max} = \frac{23}{216} \cdot \phi_{max} \cdot L^2 = \frac{23}{216} \cdot 8 \cdot 10^{-6} \cdot 240^2 \cdot \text{in.} = 0.05\ \text{ir}$$

Next, we compute the mid-span deflection for the beam under its service load. We assume the unit stress in the steel at service is

$$f_s = 36\ \text{ksi}$$

We also assume the beam to be reinforced in tension with four #8 bars and to have the following properties:

$$A_s = 4 \cdot 0.79\ \text{in.}^2 = 3.2 \cdot \text{in.}^2$$

$$f_y = 60\ \text{ksi}$$

$$f'_c = 4000\ \text{psi}$$

$$E_s - 29{,}000\ \text{ksi}$$

$$d = h - 1.5\ \text{in.} - .5\ \text{in.} = 22 \cdot \text{in.}$$

$$b = 12\ \text{in.}$$

We always check the reinforcement ratio:

$$\rho = \frac{A_s}{b \cdot d} = 1.2\%$$

For this reinforcement ratio, the depth to the neutral axis is given by

$$k = \sqrt{(\rho \cdot n)^2 + 2 \cdot \rho \cdot n} - \rho \cdot n = 0.34$$

The steel strain is (Figure 19.6d)

$$\varepsilon_s = \frac{f_s}{E_s} = 1.2 \times 10^{-3}$$

The maximum unit curvature for service loads is

$$\phi_{max} = \frac{\varepsilon_s}{d \cdot (1-k)} = 8.4 \times 10^{-5} \cdot \frac{1}{in.}$$

The deflection at mid-span is

$$\Delta_{max} = \frac{23}{216} \cdot \phi_{max} \cdot L^2 = 0.5 \text{ in.}$$

$$\frac{\Delta_{max}}{L} = 0.2\%$$

Does this result make sense? We look at it in another way:

$$\frac{L}{\Delta_{max}} = 464$$

So the calculated deflection at mid-span is approximately 1/500 of the span. It is approximately ten times the deflection at cracking. Is it a reasonable ratio? We need to think about it as we look at beams in buildings and as we calculate service load deflections for other beams.

Could we have made an educated guess? The answer is yes.

For a moderately reinforced beam we would expect the value of k to be approximately 1/3. At a reinforcement stress of 36 ksi, we can get a strain of 0.0012 by using the approximate value of 30,000 ksi. The maximum unit curvature would then be $0.0012/(2 \cdot 22/3) = 8 \cdot 10$—five or ten times the unit curvature for the uncracked section, suggesting a deflection of $10 \cdot 0.05 = ½$ in. for the service load.

It is useful to note that we did not assume a load to find the deflection. We assumed a stress in the reinforcement that corresponded to a strain. Knowing the depth to the neutral axis at service load, we could determine the maximum curvature. We used the magnitude of the curvature and its distribution along the span to determine the corresponding maximum deflection.

We need to recognize two qualifications to our calculated value:

1. We made the calculation assuming that the section was uniformly cracked throughout the span. We know that is not correct. Even in the middle third of the beam where the moment is constant, the number of cracks is bound to be finite. In between cracks, the section is likely to be stiffer. That is also true for the two outer spans where the moment decays toward the supports and cracking is even more limited. The short-time deflection calculated is likely to be higher than the actual value, assuming that the moduli are correct and that deformations caused by shear are small.
2. In an actual construction, the beam is cast in a form. The form is usually cambered based on the experience of the engineer. The visible deflection, the difference in elevation between the support and the point of maximum deflection (usually at mid-span), is likely to be smaller than the calculated value.

BARE ESSENTIALS

- Determination of the deflection of a beam is best handled in relation to its strain geometry. Knowing the distribution of the curvature along the span, we can estimate deflections.
- The simplest approach to controlling deflection is to keep the reinforcement stress low at service load and make the beam deep.

EXERCISE

Sketch the strain distribution at mid-span of beam CD of the example in Chapter 3 during service, assuming that the section is cracked and the bending moment is distributed as shown in Figure 19.7. Estimate the maximum deflection of the beam, noting that the curvature shown in Figure 19.7a is equal to the sum of those shown in Figure 19.8(b) and (c). Assume $\rho = 0.01$, $f_y = 60$ ksi, and $f_c' = 4$ ksi.

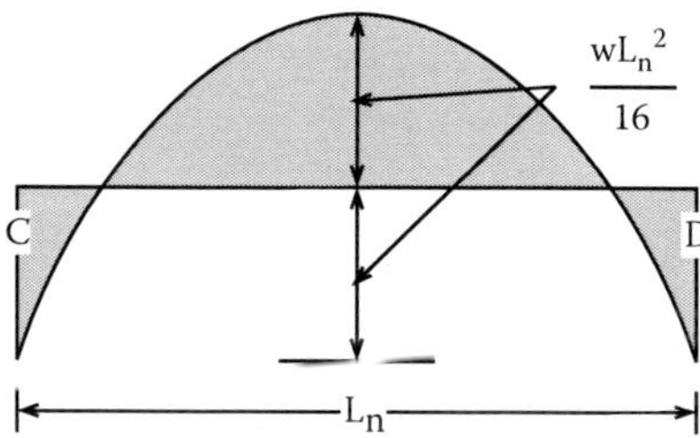

FIGURE 19.7 Curvature distribution.

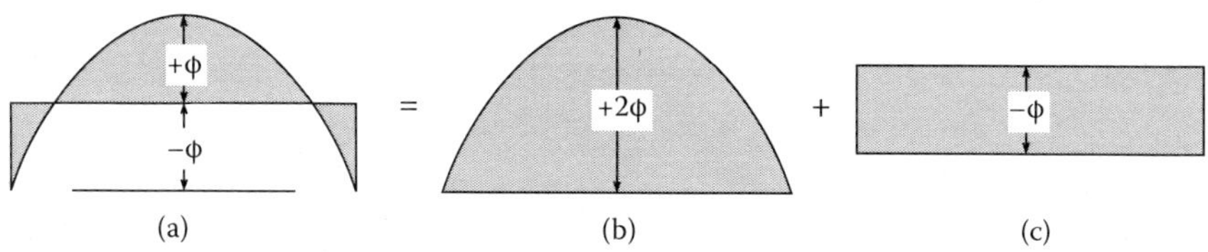

FIGURE 19.8 Decomposition of curvature.

20 Effects of Time-Dependent Variables on Deflection

As discussed in Chapter 6, time-dependent deformation of concrete is related to shrinkage and creep. Shrinkage strain is defined to occur without externally applied stress. It is expected, under typical conditions of humidity, to range from 0.0001 to as high as 0.001. The maximum is reached in approximately two years, with 80% occurring within one year (see Figure 6.6).

Creep is defined to occur as a result of externally applied stress. At applied stresses below approximately ½ of cylinder strength, the increase in strain attributed to creep is expected to be a multiple of the instantaneous strain. Its variation with time after stressing is similar to that of shrinkage. Under sustained stress, change in creep strain is negligible after two years unless there is a strong change in environmental humidity.

In the following text, we shall examine the effect of shrinkage and creep on deflections of flexural elements.

20.1 EFFECT OF SHRINKAGE

Consider a simply supported girder with the cross section shown in Figure 20.1. The section is reinforced symmetrically about its mid-height. If shrinkage occurs uniformly over the cross section or if shrinkage is symmetrical about the mid-height of the section, we may assume that no change in curvature will occur. There will be a shortening of the girder (or even lengthening if the humidity in the surrounding air is high), but there should be no associated deflection (displacement transverse to axis of girder) attributable to shrinkage.

Next consider a section reinforced in tension only (Figure 20.2). In this case, shrinkage will be restrained at the level of the reinforcement. The restraint will depend on the stiffness of the reinforcement. There will be a strain gradient over the height of the section leading to deflection away from the side of the girder with the reinforcement. In a typical continuous girder, the curvature related to shrinkage is likely to be concave up near the supports and concave down in the middle of the span because of the relative amounts of top and bottom reinforcement.

If the amount of shrinkage strain to be expected is known, there may be justification for determining by calculation the curvature caused by shrinkage in different parts of the girder. Given the curvature distribution, the shrinkage deflection may be computed directly. But knowledge of the magnitude and even sense of the shrinkage strain are not likely to be available except in very rare instances.

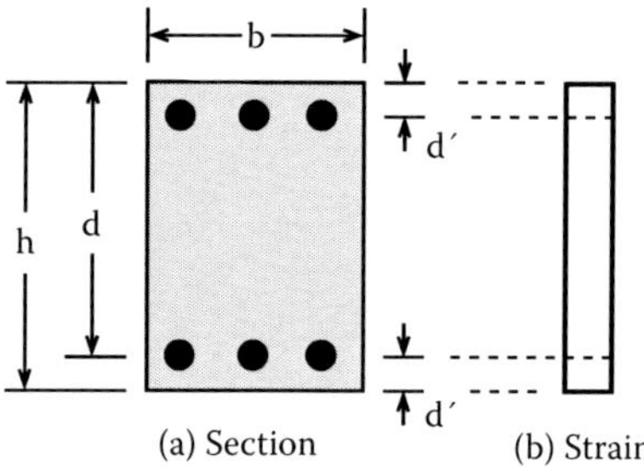

FIGURE 20.1 Symmetrically reinforced girder.

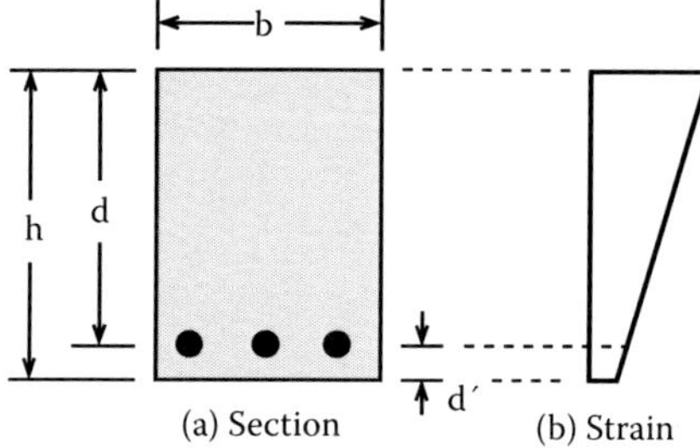

FIGURE 20.2 Girder reinforced in tension.

Should it be necessary to determine the unit curvature caused by shrinkage along the girder, the approximate expression suggested by Corley and Sozen (1966) may be used:

$$\phi_{shrink} = \frac{0.035}{d} \cdot (\rho - \rho') > 0 \tag{20.1}$$

20.2 EFFECT OF CREEP

Figure 20.3 shows the instantaneous and creep strains in compression relative to one another. If the applied stress is sustained for a very long time, and if the instantaneous strain is, say, 0.0001, the increase in strain attributed to creep is likely to be m times the instantaneous strain. The creep coefficient m is unlikely to exceed 3 (Figure 6.4) under common service conditions. At unit stresses exceeding approximately $0.6f'_c$, the creep coefficient m increases drastically with increases in unit stress.

The unit strain distributions shown in Figure 20.4 provide an idealized explanation of the fact that if the creep strain is a multiple m of the initial compressive strain, the increase in deflection will be smaller than m times the initial deflection.

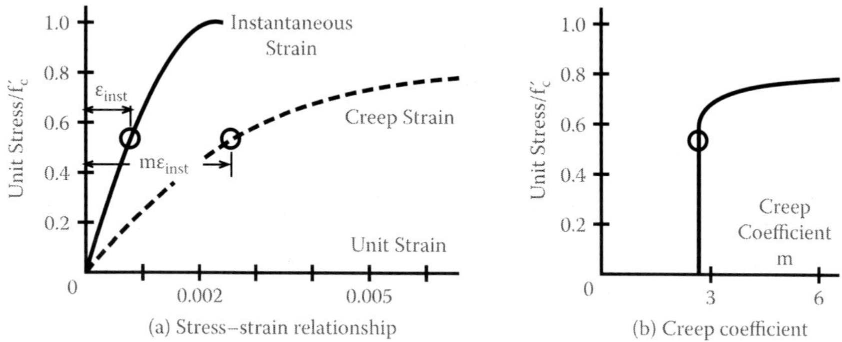

FIGURE 20.3 Instantaneous and creep strains in compression.

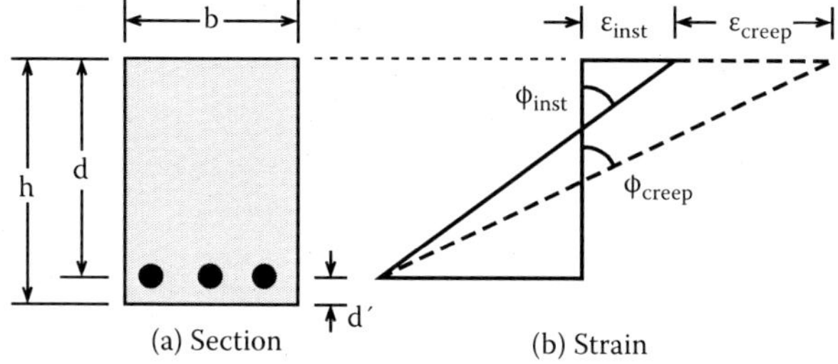

FIGURE 20.4 Strain distributions.

The solid line in Figure 20.4b represents the strain distribution occurring immediately on application of moment. Assuming that the distribution of strain over the depth of the section is linear, the initial unit curvature is

$$\phi_{inst} = \frac{\varepsilon_{inst}}{k \cdot d} \tag{20.2}$$

where $k \cdot d$ is the depth to the neutral axis from the extreme fiber in compression. The instantaneous deflection is proportional to ϕ_{inst} in a beam with reasonably uniform flexural cracking.

If the moment remains on the section, the increase in maximum strain can be assumed to be $m \cdot \varepsilon_{inst}$.

If the strain distribution remains linear, the depth of the neutral axis increases. The concrete stress is reduced because it is spread over more of the depth, and to maintain the moment, the steel strain increases because the internal lever arm, jd, decreases. To obtain a pragmatic estimate of the increase in unit curvature, we can ignore the change in steel strain. As represented by the dashed line in Figure 20.4b, the increase in unit curvature ϕ_{creep} becomes

$$\phi_{creep} = \frac{m \cdot \varepsilon_{inst}}{d} \tag{20.3}$$

leading to

$$\frac{\phi_{creep}}{\phi_{inst}} = k \cdot m \tag{20.4}$$

The result above contains the most revealing statement about the influence of creep strain on deflection. The creep strain may be m times the instantaneous strain, but the creep curvature (and therefore the deflection attributable to creep) is only $k \cdot m$ times the instantaneous deflection. For sections used generally in beams, the value k ranges from 0.3 to 0.4. If m = 3, creep deflection would be expected to be of the same magnitude as the instantaneous deflection.

If compression reinforcement is used, increase in compressive strain at the level of the compression reinforcement leads to an increase in compressive stress in the reinforcement if it has not already yielded. Because the force in the tension reinforcement tends to remain at the same level, the stress in the compressed concrete reduces at a faster rate. The reduction in compressive stress reduces the magnitude of the creep strain. This complex interrelationship was represented by Corley and Sozen using a very simple relationship:

$$\phi_{creep} = \left(3 - \frac{\rho'}{\rho}\right) \cdot k \cdot \phi_{inst} \tag{20.5}$$

The curvature related to creep as defined above refers to a best estimate for normal weight aggregate concrete. In using this expression, it must be kept in mind that the actual change in curvature related to creep may be higher or lower by 50%. In dealing with time-dependent effects, one does not seek accuracy.

EXAMPLE 20.1

Because of their industrial function, pipes are to be connected at mid-spans of simply supported precast reinforced concrete girders spanning 26 ft, as shown in Figure 20.5. The girder section is shown in Figure 20.6.

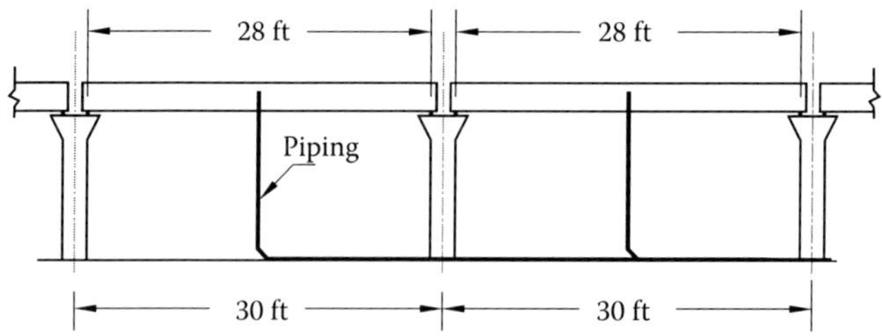

FIGURE 20.5 Girder and pipe layout.

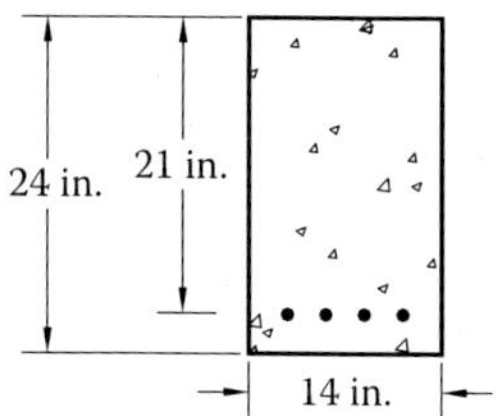

FIGURE 20.6 Girder section.

Compressive strength of concrete $f'_c = 4000$ psi

Yield stress of reinforcement $f_y = 60$ ksi

Young's modulus of concrete $E_c = 57{,}000 \cdot \sqrt{f'_c \cdot psi} = 3.6 \text{ x } 103$ ksi

Young's modulus of steel $E_s = 29{,}000 \text{ ksi} = 2.9 \times 104$ ksi

Modular ratio $n = E_s \div E_c = 8.0$

Overall depth of section $h = 24$ in.

Effective depth of section $d = 21$ in.

Width $b = 14$ in.

Span $L = 26$ ft

Area of tension reinforcement $A_s = 4 \text{ in.}^2$

Tension on reinforcement ratio $\rho = \dfrac{A_s}{b \cdot d} = 1.4\%$

Loads:

Tributary dead load $w_{DL} = 1 \text{ kip} \div \text{ft}$

Tributary live load

Sustained $w_{SLL} = 0.6 \text{ kip} \div \text{ft}$

Transient $w_{TLL} = 0.6 \text{ kip} \div \text{ft}$

Service-load maximum moment $M_{service} = (w_{DL} + w_{SLL} + w_{TLL}) \cdot L2 \div 8$

$M_{service} = 186$ kip-ft

Sustained maximum moment $M_{sustained} = (w_{DL} + w_{SLL}) \cdot L^2 \div 8$

$M_{sustained} = 135$ kip-ft

Moment at cracking $M_{cr} = b \cdot h^2 \sqrt{f'_c \cdot psi} = 42.5$ kip-ft

Depth to neutral axis in linear range of response (after cracking):

$$k = \sqrt{\rho^2 \cdot n^2 + 2 \cdot \rho \cdot n} - \rho \cdot n$$

$$k = 0.37$$

Service-load stresses in the reinforcement:

For total loads (sustained and transient)

$$f_s = \frac{M_{service}}{A_s \cdot \left(1 - \frac{k}{3}\right) \cdot d}$$

$$f_s = 30 \text{ ksi}$$

For sustained loads $f_{s\_sustained} = \dfrac{M_{sustained}}{A_s \cdot \left(1 - \dfrac{k}{3}\right) \cdot d}$

$f_{s\_sustained} = 22 \text{ ksi}$

Service-load strains in the reinforcement:

For total loads (sustained and transient) $\varepsilon_s = f_s \div E_s$

$\varepsilon_s = 1 \times 10^{-3}$

For sustained loads $\varepsilon_{inst} = f_{s\_sustained} \div E_s$

$\varepsilon_{inst} = 7.6 \times 10^{-4}$

Instantaneous service-load unit curvatures (Figure 20.7):

For total loads (sustained and transient) $\phi = \varepsilon_s \div (d - k \cdot d)$

$\phi = 7.9 \times 10^{-5} \dfrac{1}{\text{in.}}$

$\phi_{inst} = \varepsilon_{inst} \div (d - k \cdot d)$

$\phi_{inst} = 5.8 \times 10^{-5} \dfrac{1}{\text{in.}}$

Instantaneous service-load deflections:

For total laods (sustained and transient) $\Delta = \dfrac{5}{48} \cdot \phi \cdot L^2$

$\Delta = 0.8 \text{ in.}$

For sustained loads $\Delta_{inst} = \dfrac{5}{48} \cdot \phi_{inst} \cdot L^2$

$\Delta_{inst} = 0.6 \text{ in.}$

Estimated time-dependent deflections:

Related to shrinkage $\rho' = 0$

$\phi_{shrink} = \dfrac{0.035}{d}(\rho - \rho') = 2.3 \times 10^{-5} \dfrac{1}{\text{in.}}$

$\Delta_{shrink} = \dfrac{1}{8} \cdot \phi_{shrink} \cdot L^2$

$\Delta_{shrink} = 0.3 \text{ in.}$

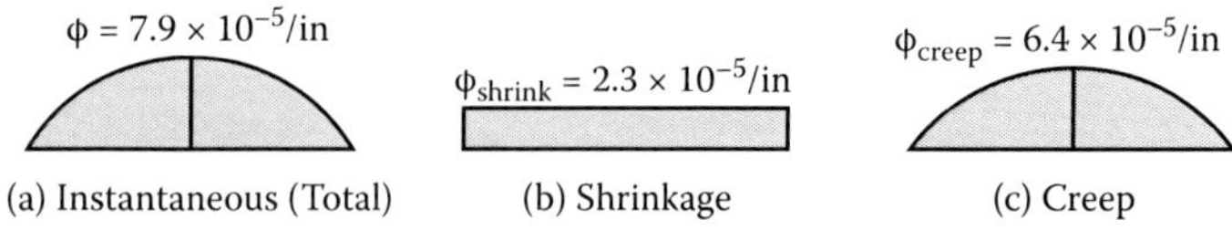

FIGURE 20.7 Assumed curvature distribution.

Related to creep

$$\phi_{creep} = \left(3 - \frac{\rho'}{\rho}\right) \cdot k \cdot \phi_{inst}$$

$$\Delta_{creep} = \frac{5}{48} \cdot \phi_{creep} \cdot L^2$$

$$\Delta_{creep} = 0.6 \text{ in.}$$

$$\text{Note: } \left(3 - \frac{\rho'}{\rho}\right) \cdot k \cdot \Delta_{inst} = 0.6 \text{ in.}$$

Total deflection

$$\Delta_{total} = \Delta + \Delta_{shrink} + \Delta_{creep}$$

$$\Delta_{total} = 1.7 \text{ in.}$$

$$\frac{L}{\Delta_{total}} = 181$$

Estimate the maximum lateral deflection of the pipe:

Height

$$H = 12 \text{ ft}$$

$$H' = H - \Delta_{total}$$

Assuming a triangular shape

$$x_{max} = \sqrt{\left(\frac{H}{2}\right)^2 - \left(\frac{H'}{2}\right)^2}$$

$$x_{max} = 11 \text{ in.}$$

Assuming a sinusoidal deformed shape does not change the result much (the broken line in Figure 20.8), the lateral displacement of the pipe is unacceptable.

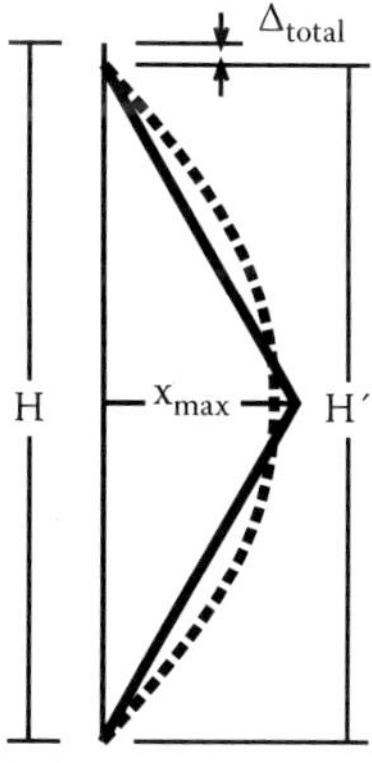

FIGURE 20.8 Deformed shape of pipe in Example 20.1.

BARE ESSENTIALS

Should it be necessary to estimate the unit curvature caused by shrinkage and creep, the following expressions may be used:

$$\phi_{shrink} = \frac{0.035}{d} \cdot (\rho - \rho') > 0$$

$$\phi_{creep} = \left(3 - \frac{\rho'}{\rho}\right) \cdot k \cdot \phi_{.inst}$$

EXERCISE

To try to solve the problem of intolerable estimated deflection described above, go to an extreme: increase the overall depth of the girder to 48 in. Over a span of 26 ft, the depth is unquestionably unreasonable, but try that to see whether the lateral deflection of the piping can be reduced so that it is not obvious to the human eye. Also add compression reinforcement to be provided by four # 9 bars.

Note: The depth to the neutral axis for a cracked section with compression reinforcement is given by

$$k = \sqrt{\beta^2 + 2\rho n + 2\rho' n \gamma_c} - \beta$$

where $\beta = \rho \cdot n + \rho' n$. The student is encouraged to derive this expression on his or her own.

Recall

$$\gamma_c = d'/d$$

where d′ is the depth to compression reinforcement, and

$$\rho' = A'_s/(bd)$$

where A'_s is the cross-sectional area of compression reinforcement (notice that d—not d′—is used in this definition).

21 Continuous Beams

Despite advances in applications of limit analysis and despite recognition of the effects on stiffness of time-dependent variables, determination of moment demands for continuous beams is driven by two criteria that date from the time of working stress design: (1) moments in a reinforced concrete structure can be determined accurately on the basis of linear response, and (2) moments so determined must not be exceeded. Successful use of a design procedure based on those criteria provides another example of a design method that defies observed behavior but leads to acceptable results. Referring to a beam restrained at both ends, we develop a perspective of how this contradiction in concept leads to tolerable results.

If we ignore the effect of cracking in a uniformly loaded beam with ends fixed against rotation and consider it to be prismatic (uniform bending stiffness throughout the span), the ratio of the moment at the support, M_A, to that at mid-span, M_B, will be determined to be $M_A/M_B = 2/1$ (Figure 21.1a and b). However, sections at support and at mid-span are proportioned typically to have similar, if not the same, steel stresses under service load (ε_A and ε_B in Figure 21.1c and d). If they have similar steel stresses, they will have similar strains and unit curvatures (ϕ_A and ϕ_B in Figure 21.1e). That condition defeats any argument for a prismatic beam. In a prismatic beam, the unit curvature at the supports should be twice that at mid-span, so that the slope of the beam is zero at the ends and at mid-span.

The selection of equal steel stresses at all critical sections contradicts the assumption of the unit curvature varying as it would for a prismatic beam. What are the effects of this contradiction on the strength and deformations of the beam? To answer the question, we refer to a hypothetical beam (Figure 21.2).

Consider the uniformly loaded prismatic beam shown in Figure 21.2a. Its ends are restrained against rotation by linear springs. We know from applications of the theory of elasticity that if the springs are removed, the moments at the support and mid-span are (Figure 21.2b)

$$M_{support} = 0 \tag{21.1}$$

$$M_{midspan} = \frac{wL^2}{8} \tag{21.2}$$

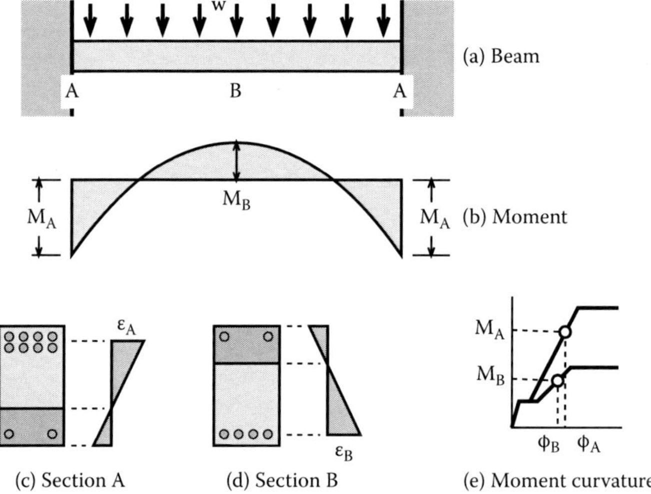

FIGURE 21.1 Idealized moments, strains, and unit curvatures in a beam designed to have similar steel stresses at supports and at mid-span.

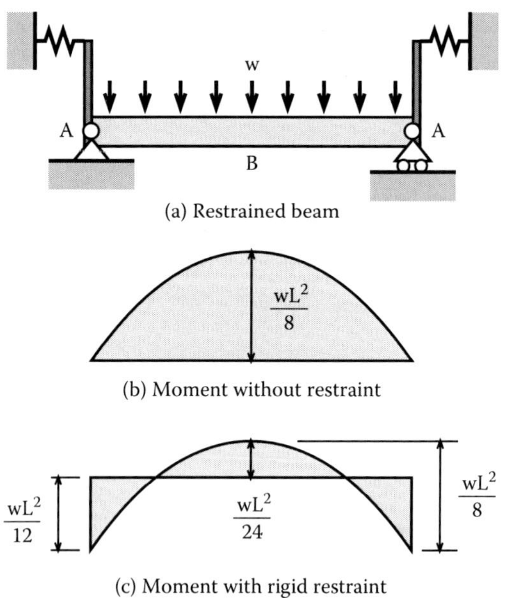

FIGURE 21.2 A hypothetical beam with ends restrained against rotation by linear springs.

If the springs are rigid (Figure 21.2c),

$$M_{support} = \frac{wL^2}{12} \tag{21.3}$$

$$M_{midspan} = \frac{wL^2}{24} \tag{21.4}$$

We also note that the absolute sum of the moments at support and at mid-span is

$$M_0 = M_{support} + M_{midspan} = \frac{wL^2}{8} \tag{21.5}$$

We recall the definition of M_0 to be the static moment obtained directly from statics. Static equilibrium in a uniformly loaded beam restrained equally at each end requires that the sum of the moments at one support and at mid-span must be equal to $wL^2/8$, or to the simple beam moment. This condition holds for all symmetrically loaded beams. For example, for a beam of length L with a concentrated load, P, at its mid-span, the static moment is PL/4. If both ends of this beam are fixed against rotation, the absolute sum of one support moment plus the moment at mid-span is equal to PL/4.

It follows from the discussion above that depending on the rotation restraints at the supports, the absolute ratio of the negative moment at the support to the positive moment at mid-span may vary from 0 to 2, but the static moment remains constant.

Hardy Cross* developed an ingenious method, the column analogy, for determining fixed-end moments in single-span beams and frames. Using the simplest application of this method, we try to develop a perspective of the effect of variation of section stiffness along the span of a symmetrical and symmetrically loaded beam.

His proposed method is simply the following. Assume that the beam is analogous to a section with height equal to the span of the beam and its width defined as 1/EI, where E is the Young's modulus for the beam material and I is its moment of inertia (Figure 21.3). For that beam, the section area of the analogous section would be simply

$$A_a = \frac{L}{EI} \tag{21.6}$$

where A_a is the cross-sectional area of the analogous section, L is the span of the beam assumed to be the height of the analogous section, and EI is the product of Young's modulus for the beam material and the moment of inertia of the beam section assumed to be the inverse of the thickness of the analogous section.

* Cross, H., *The Column Analogy*, University of Illinois Engineering Experiment Station Bulletin 215, October 1930.

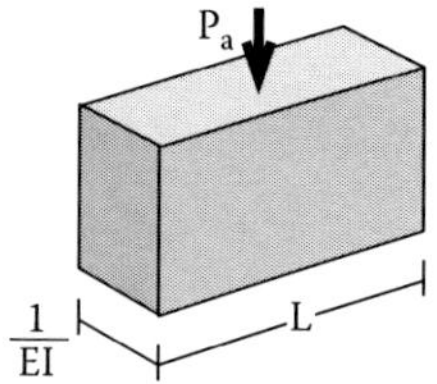

FIGURE 21.3 Analogous column.

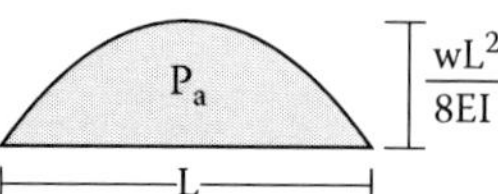

FIGURE 21.4 Analogous load.

The fixed-end moments, or the support moments, in a beam with ends fully restrained against rotation are determined as the unit stress caused on the analogous section by an axial load equal to the area of the diagram of the unit curvature that the applied load would cause in a simply supported beam.

We know the maximum simple beam moment for a uniformly loaded beam is

$$M_0 = \frac{wL^2}{8} \tag{21.7}$$

We also know that it varies parabolically from zero to a maximum at mid-span. Therefore, the analogous load is the area of the parabola shown in Figure 21.4, or

$$P_a = \frac{2}{3}\frac{wL^2}{8}\frac{L}{EI} = \frac{wL^3}{12EI} \tag{21.8}$$

Dividing the analogous load by the analogous section area, we arrive at the moment at the support,

$$\frac{P_a}{A_a} = \frac{\dfrac{wL^3}{12EI}}{\dfrac{L}{EI}} = \frac{wL^2}{12} \tag{21.9}$$

a quantity that does not surprise us (Figure 21.2c). The fixed-end moment amounts to 2/3 of the static moment in a prismatic uniformly loaded beam.

Now we go further and ask, how would the support moment vary if the beam was nonprismatic (Figure 21.5a)?

We ask that question because we know that typically the section at the support of a reinforced concrete beam has more longitudinal reinforcement than the section at mid-span. Once the beam cracks, the stiffness of the sections with negative moment tends to be higher than that of sections with positive moment because of the relative amounts of reinforcement.

We should be able to get a compact answer to that question using the column analogy method.

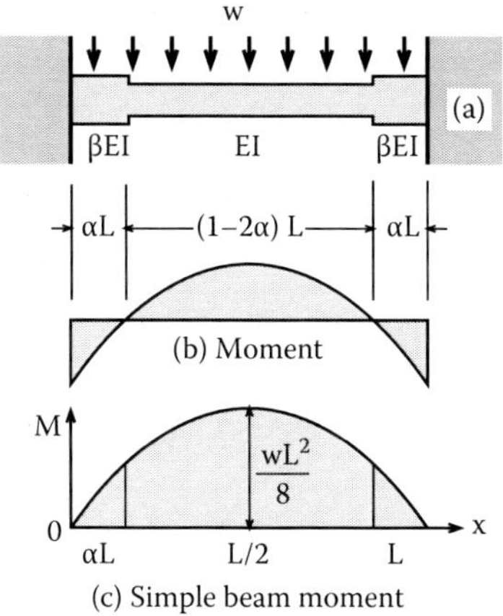

FIGURE 21.5 Nonprismatic beam.

The simple beam moment varies with distance from the support, x, as shown in Figure 21.5c, or

$$M = \frac{w}{2}\left[\frac{x}{L} - \left(\frac{x}{L}\right)^2\right] \tag{21.10}$$

Let us assume that the portion of the beam subjected to positive moment has a section stiffness EI, and the portions subjected to negative moment have section stiffnesses βEI, where β is more than 1 (Figure 21.5a and b).

The analogous load (Figure 21.6) is expressed in two parts. Within the negative moment span it is

$$P_n = \int_0^{\alpha L} \frac{M}{\beta EI} dx = \frac{w}{2\beta EI} \int_0^{\alpha L} \left[\frac{x}{L} - \left(\frac{x}{L}\right)^2\right] dx \tag{21.11}$$

Within the positive moment span, it is

$$P_p = \int_{\alpha L}^{L/2} \frac{M}{EI} dx = \frac{w}{2EI} \int_{\alpha L}^{L/2} \left[\frac{x}{L} - \left(\frac{x}{L}\right)^2\right] dx \tag{21.12}$$

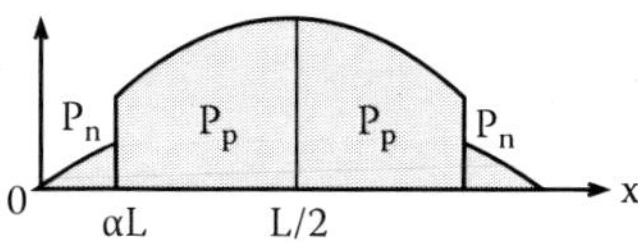

FIGURE 21.6 Analogous load.

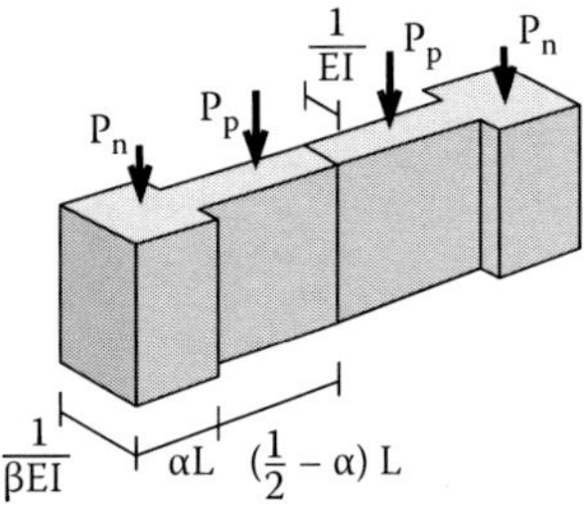

FIGURE 21.7 Analogous column.

The cross-sectional area of the analogous column (Figure 21.7) is

$$A_a = 2\left(A_n + A_p\right) = 2\left[\frac{1}{\beta EI} \times \alpha L + \frac{1}{EI} \times \left(\frac{1}{2} - \alpha\right) L\right] \tag{21.13}$$

The unit stress on the section or the fixed-end moment for the beam is

$$\text{FEM}\left(\alpha,\beta\right) = \frac{2\left(P_n + P_p\right)}{A_a} = \frac{P_n + P_p}{A_n + A_p} \tag{21.14}$$

In Figure 21.8 we examine the variation of the FEM with changes in α and β. We note the following trends:

For β = 1 (prismatic beam), the FEM is constant at $wL^2/12$, or 2/3 of the static moment.

For β = 2, the coefficient for wL^2 varies from 1/12 at α = 0 to a value approximately 15% higher in the range α = 0.2 to α = 0.3. At α = 0.5, the FEM is the same as it was at α = 0. That result does not surprise us. If α = 0.5, the beam is prismatic.

For β = 4, the maximum occurs at approximately 0.3, where the increase in FEM approaches 25% of its value at α = 0.

From the data in the plot and from our judgment that the negative moment section is not likely to be more than twice as stiff as the positive moment section, we infer that we need not be concerned about the increase in negative moment caused by cracking. Our conclusion is strengthened by recognizing that under practical conditions, the ends of the beam are not fully restrained. A small rotation at the end would reduce the negative moment. It is very difficult to establish the effects of changes in stiffness and end rotation by calculation. No matter what we assume, we are likely to be wrong. In that case, we assume the beam to be prismatic. As the old engineering adage goes, as long as we are going to be wrong, we should be wrong the easy way.

So we go ahead and "break" a beam with fixed ends into two cantilever beams and a simple beam, as shown in Figure 21.9.

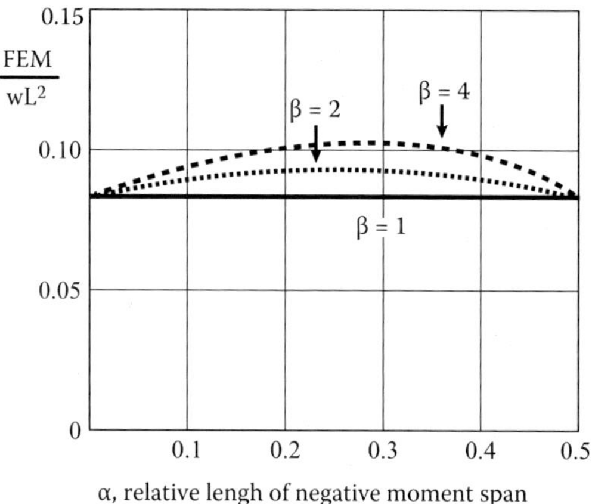

FIGURE 21.8 Variation of FEM with changes in α and β.

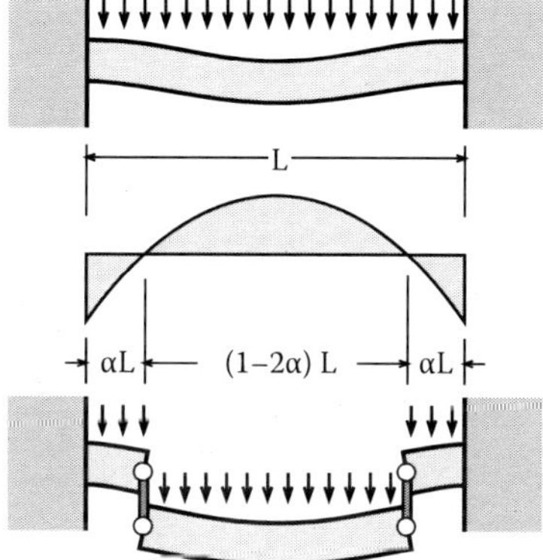

FIGURE 21.9 A beam with fixed ends represented by two cantilever beams and a simple beam.

To determine the lengths of the cantilevers, we assume a moment distribution based on a prismatic beam with fixed ends (support moment = $wL^2/12$) and seek the position where moment is zero

$$\frac{L^2}{12} - \frac{L}{2}\alpha L + \frac{(\alpha L)^2}{2} = 0 \tag{21.15}$$

leading to the solutions

$$\alpha = \frac{1}{2} \pm \frac{\sqrt{3}}{6} \tag{21.16}$$

So we assume that

$$\alpha = \frac{1}{2} - \frac{\sqrt{3}}{6} \approx 0.21 \tag{21.17}$$

in determining the deflections of a beam with fixed ends, which can be computed as the sum of the deflection of a cantilever beam and the deflection of a simply supported beam (Figure 21.9).

EXAMPLE 21.1

A beam with ends restrained against rotation has a clear span L = 21 ft. It resists a load of 30 kip applied at a distance a = 7 ft from one support. Compute the end moments neglecting the weight of the beam and assuming:

1. The beam is prismatic (Figure 21.10a).
2. The flexural stiffness of the outer thirds of the span is two times the stiffness of the middle third (Figure 21.10b).

SOLUTION

Define the distance, a, from the support closer to the applied load as x and define b = L − a = 14 ft (Figure 21.11a). The moment distribution that the applied load would cause in a simply supported beam is (Figure 21.11b)

$$M(x) = P \cdot \frac{b}{L} \cdot a \begin{vmatrix} \frac{x}{a} & \text{if } x < a \\ \frac{1}{b} \cdot (L - x) & \text{otherwise} \end{vmatrix}$$

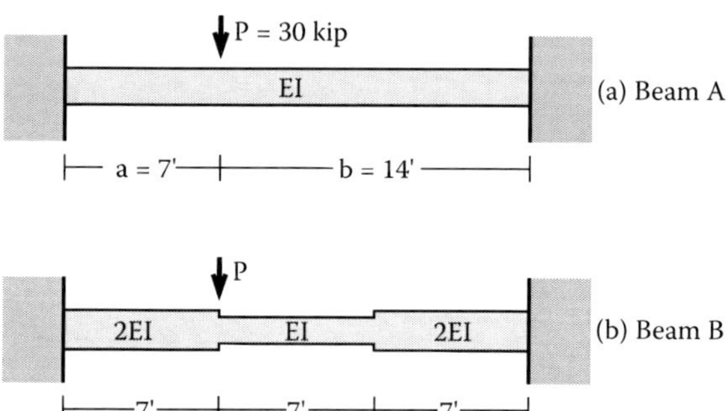

FIGURE 21.10 Prismatic and nonprismatic beams.

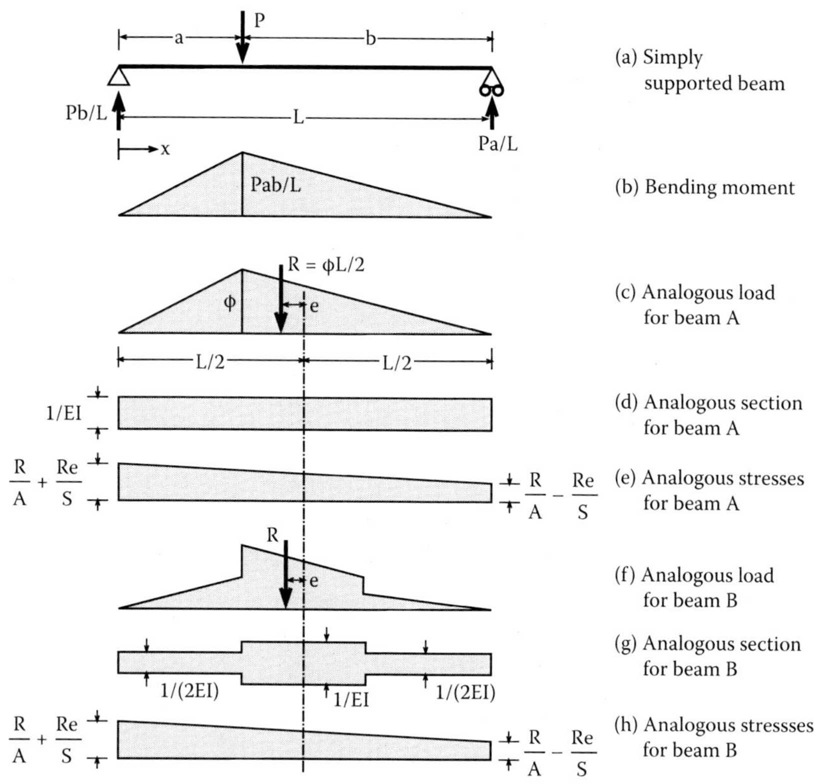

FIGURE 21.11 Analogous sections and stresses.

Defining the relative stiffness of the cross section as EI(x), the analogous load is:	$R = \int_0^L \frac{M(x)}{EI(x)}\,dx$	Figure 21.11b
The eccentricity of the analogous load, relative to mid-span:	$e = \frac{L}{2} - \frac{1}{R} \cdot \int_0^L x \cdot \frac{M(x)}{EI(x)}\,dx$	Figure 21.11c or f
The area of the analogous section:	$A = \int_0^L \frac{1}{EI(x)}\,dx$	Figure 21.11d or g
The analogous section modulus:	$S = \frac{2}{L} \cdot \left[\int_0^L \frac{1}{EI(x)} x^2\,dx - A \cdot \left(\frac{1}{2} \cdot L\right)^2\right]$	Figure 21.11d or g
The analogous stresses (end moments):	$R/A \pm R \cdot e/S$	Figure 21.11e or h

With the help of software for numerical integration and defining the relative sectional stiffness as (Case 1) EI = 1 and (Case 2).

$$EI(x) = \begin{vmatrix} 1 \text{ if } \frac{L}{3} < x < \frac{2}{3} \cdot L \\ 2 \text{ otherwise} \end{vmatrix},$$

we obtain the following results

Case	R (kip-ft²)	e (ft)	A (ft)	S (ft²)	R/A + R·e/S	R/A - R·e/S
1	1470	1.2	21	73	0.15 PL	0.07 PL
2	1103	0.91	14	38	0.17 PL	0.08 PL

Note: To appreciate how efficient the column analogy is, consider the following alternate solution for Case 1 based on the moment-area theorems: First, consider a simply supported beam carrying the same load as the given beam (Figure 21.12a). Its slope at the support closer to the applied load θ_A is equal to the first moment of the area of the curvature diagram taken about the opposite support (t_{BA}) divided by the span (Figure 21.12b and c):

$$\theta_A = \frac{t_{BA}}{L} = \left[\frac{1}{2}\cdot\left(\frac{2}{3}\cdot L\cdot\varphi_{max}\right)\cdot\left[\frac{2}{3}\cdot\left(\frac{2}{3}\cdot L\right)\right] + \frac{1}{2}\cdot\left(\frac{1}{3}\cdot L\cdot\varphi_{max}\right)\cdot\left[\frac{1}{3}\left(\frac{1}{3}\cdot L\right) + \frac{2}{3}\cdot L\right]\right] \div L$$

$$= \frac{5}{18}\cdot\phi_{max}\cdot L$$

where ϕ_{max} is the maximum curvature 2 PL/(9 EI). Similarly, the slope at the other support θ_B is

$$\theta_B = \frac{t_{AB}}{L} = \left[\frac{1}{2}\cdot\left(\frac{2}{3}\cdot L\cdot\varphi_{max}\right)\cdot\left[\frac{1}{3}\cdot\left(\frac{2}{3}\cdot L\right) + \frac{1}{3}\cdot L\right] + \frac{1}{2}\cdot\left(\frac{1}{3}\cdot L\cdot\varphi_{max}\right)\cdot\left[\frac{2}{3}\cdot\left(\frac{1}{3}\cdot L\right)\right]\right] \div L$$

$$= \frac{2}{9}\cdot\varphi_{max}\cdot L$$

These slopes are equal and opposite to the slopes caused by the end moments. Each end moment causes the following slopes:

$$\text{Where it acts: } \left(\frac{2}{3}\right)\cdot\left(\frac{1}{2}\cdot\frac{M_{end}}{EI}\cdot L\right)$$

$$\text{At the opposite support: } \left(\frac{1}{3}\right)\cdot\left(\frac{1}{2}\cdot\frac{M_{end}}{EI}\cdot L\right)$$

Defining M_A as the end moment closer to the applied load and M_B as the moment at the opposite support (Figure 21.12d), and superimposing the effects of the applied load and the end moments (Figure 21.12e):

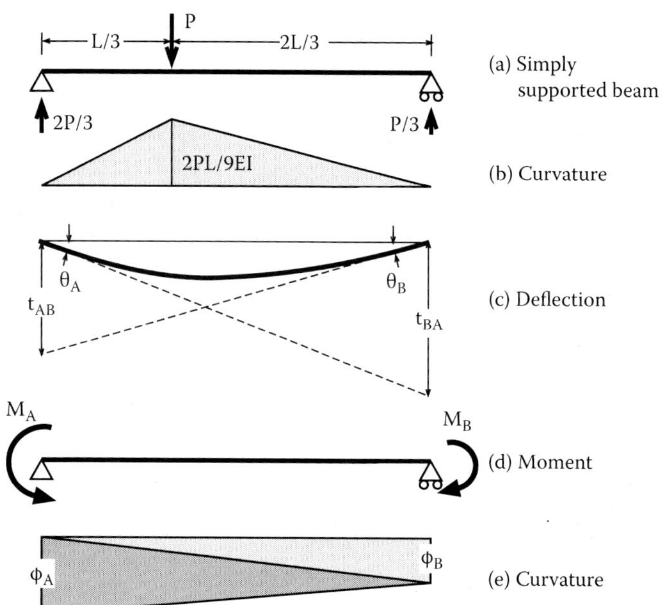

FIGURE 21.12 Solution based on moment–area theorem.

$$\frac{2}{3}\left(\frac{1}{2}\cdot\frac{M_A}{EI}\cdot L\right)+\frac{1}{3}\cdot\left(\frac{1}{2}\cdot\frac{M_B}{EI}\cdot L\right)=\frac{5}{18}\cdot\left(\frac{2}{9}\cdot\frac{P\cdot L}{EI}\right)\cdot L$$

$$\frac{1}{3}\left(\frac{1}{2}\cdot\frac{M_A}{EI}\cdot L\right)+\frac{2}{3}\cdot\left(\frac{1}{2}\cdot\frac{M_B}{EI}\cdot L\right)=\frac{2}{9}\cdot\left(\frac{2}{9}\cdot\frac{P\cdot L}{EI}\right)\cdot l$$

Simplifying and rewriting these equations:

$$\begin{pmatrix} 1 & \frac{1}{2} \\ \frac{1}{2} & 1 \end{pmatrix}\cdot\begin{pmatrix} M_A \\ M_B \end{pmatrix}=\frac{2}{27}\cdot\begin{pmatrix} \frac{5}{2} \\ 2 \end{pmatrix}\cdot P\cdot L$$

In this system of equations, the left-hand side is independent of the loading and remains the same as long as the beam is prismatic. But the right-hand side does depend on both load and stiffness distributions. The right-hand side can be written in more general terms using integrals to compute the first moments of the curvature diagram with respect to the supports. For prismatic beams, the main advantage of the column analogy is that it does not require the solution of a system of simultaneous equations.

Solving the system of equations obtained leads to the same answers computed using the column analogy: $M_A = 0.15$ PL and $M_B = 0.07$ PL. The advantage of the moment-area method is that the engineer using it is likely to understand the behavior of the beam under the applied loading. However, the ratio of the result to labor is low, and the likelihood of making an error in setting up the equations is high. The advantage of the column analogy is that the ratio of the result to labor is very high.

BARE ESSENTIALS

- The static moment in a single span is closely equal to the absolute sum of the mean of the support moments and the moment at mid-span.
- In symmetrically reinforced, loaded, and restrained beams, the absolute sum of one support moment and the moment at mid-span is exactly equal to the static moment.
- In restrained beams, the ratio of the support moment to the positive moment is not sensitive to the effect of reinforcement distribution along the span. Treating the beam as a prismatic beam is plausible.
- Because of cracking and time-dependent effects, we should always be skeptical about the accuracy of moment distributions determined by linear analysis in continuous elements.

EXERCISE

Compute the end moments for a *prismatic* beam with

1. Supports restrained against rotation.
2. A clear span of 25 ft.
3. A triangular load distribution with a maximum load of 1 kip-ft at mid-span.

Ignore the weight of the beam.

22 Limiting Load

The quantitative relationship between moment and unit curvature for an under-reinforced section was developed in Chapter 14 (Figure 14.1). The shape of the curve suggests that the section is likely to have considerable rotation capacity after yielding. If this is true for the critical sections of a continuous beam, yielding at one or even two sections does not limit the load capacity of the beam.

To understand the phenomenon of a continuous beam continuing to resist increases in load after yielding of one of its sections, let us consider a beam in a frame (Figure 22.1a) and assume a rather unusual arrangement of reinforcement: a beam reinforced equally at top and bottom (Figure 22.1b), with both sets of bars extending throughout the span and anchored effectively in adjoining members of the frame.

The reinforcement arrangement we have assumed would lead to equal moment capacities at all sections of a prismatic beam with rectangular section, whether the section is subjected to a positive (tension at bottom) or negative (tension at top) moment.

If the beam shown in Figure 22.1a is loaded uniformly along all of its spans, and if we may, for the sake of simplicity, assume that its bending stiffness is uniform throughout its span, we can conclude that initially the ratio of the moment demand magnitudes at the supports (at the face of the columns) to that at mid-span is close to 2 (or –2 if one wants to stick to the sign convention). As shown in Figure 22.1c, this distribution of moment demand would result in yielding of the sections at the supports first. At that instant, the moment at mid-span would be half that at the support. Considering one-half of the beam (Figure 22.1d), we can set up an equation based on statics as follows:

$$w \cdot \frac{L}{2} \cdot \frac{L}{4} = M_{support} + M_{midspan} \tag{22.1}$$

where w is the unit load, L is the clear span, $M_{support}$ is the moment at the face of support equal to the yield moment, M_y, and $M_{mid\text{-}span}$ is the moment at mid-span equal to one-half of the moment at support.

Equation 22.1 may be rewritten as

$$\frac{w \cdot L^2}{8} = \frac{3}{2} \cdot M_y \tag{22.2}$$

For a symmetrical arrangement of moment demands and resistances, Equation 22.1 makes a very simple statement: The simple beam moment (also known as the static moment) is equal to the absolute sum of the moment resistances at the support and at mid-span. This simple statement helps us anticipate the events at the end of the next step in loading. As the load is increased (Figure 22.2a), the moment at the support remains

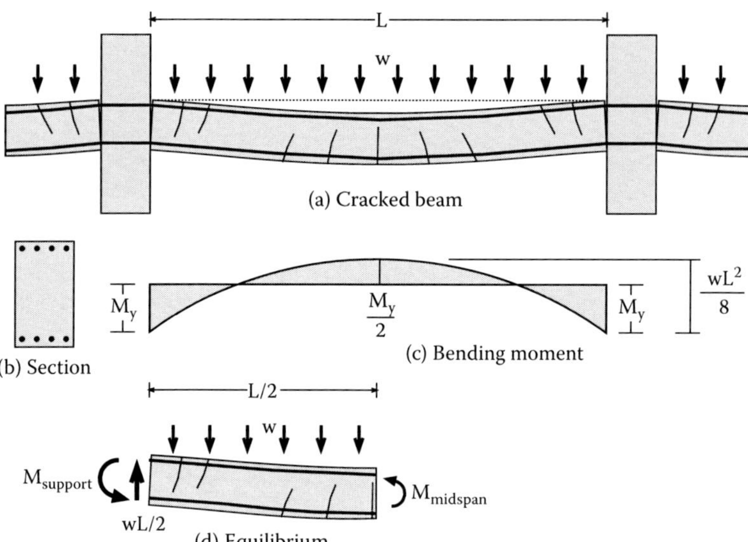

FIGURE 22.1 Interior beam at elastic limit.

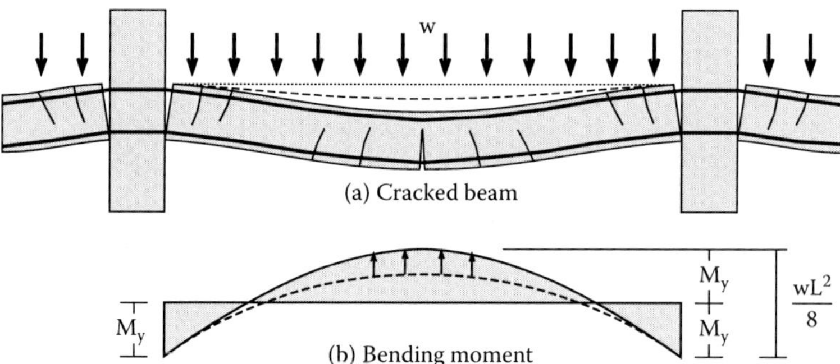

FIGURE 22.2 Interior beam under limiting load.

the same, but the moment at mid-span increases to its yield value M_y (Figure 22.2b). Once that moment level is reached at mid-span, the beam will continue to deform, but the moments at supports and at mid-span will remain constant. Therefore, Equation 22.1 can be written as

$$\frac{w \cdot L^2}{8} = 2 \cdot M_y \tag{22.3}$$

In this case, we have simplified the problem by assuming the reinforcement to be the same at the top and bottom, and constant along the span. The reinforcement arrangement in a representative interior beam is likely to look as shown in Figure 22.3.

Usually the reinforcement at the support is discontinued, if only gravity load is considered, so that at L/3 from the face of support (where L is the clear span) all

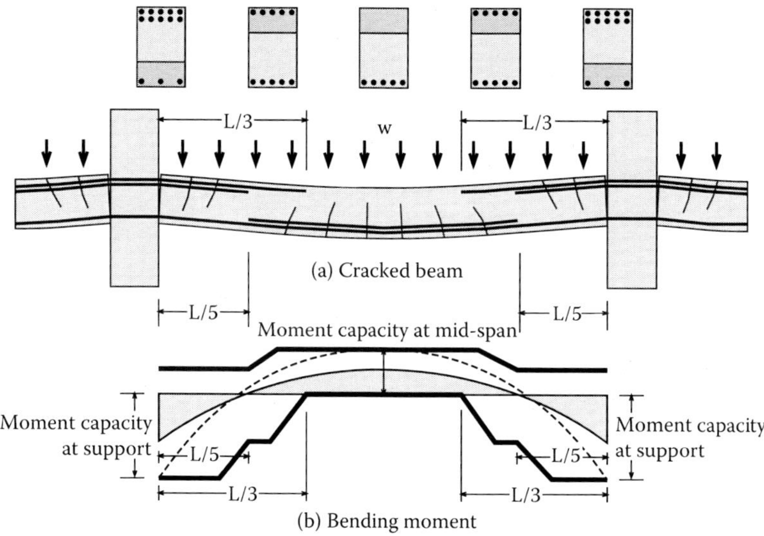

FIGURE 22.3 Reinforcement in a representative interior beam.

the top reinforcement is cut off, while at L/5 from the support half of it may be cut off. Similarly, half the bottom reinforcement may be cut off at a distance L/5 from the support. These lengths are typically governed by the applicable building code, although, very often, the engineer selects them to be more than that required.

With the arrangement of reinforcement shown in Figure 22.3a, the resisting moment varies along the span. Assuming that bond stress is distributed uniformly (see Chapter 27), the thick lines in Figure 22.3b delineate the boundaries of the limiting moment capacity.

As the continuous beam is loaded (assuming that the adjacent beams are loaded at the same rate), the support moment is nearly twice that at mid-span. In this example we assume that the support moment capacity is three times that at mid-span, M_y. Accordingly, the section at mid-span yields first. The load continues to increase, with the sections at the supports resisting all of the increase until they yield. Under those conditions, Equation 22.2 enables determination of the maximum unit load. If the static moment, $wL^2/8$, is equal to the absolute sum of the moments at the critical sections,

$$\frac{w \cdot L^2}{8} = M_{support} + M_{midspan} = 3 \cdot M_y + M_y = 4 \cdot M_y \tag{22.4}$$

It is important to note that, in this case, with a nonuniform distribution of reinforcement along the span, the moment demand diagram along the span must always remain within the capacity limits indicated by the thick lines in Figure 22.3. That is the reason for experienced engineers electing to cut off the bars at longer lengths than lengths that may be rationalized by the distribution of the moment demand diagram based on linear analysis. It should also be noted that the degree of restraint at the supports (because of the stiffness and loading of the adjoining spans) may also affect the distribution of the moment within the span considered.

EXAMPLE 22.1

A beam has a clear span L = 21 ft, and its ends are restrained against rotation. It supports a concentrated load P applied at 7 ft from one of the supports.

Assume the following:

1. The cross section (Figure 22.4) is constant (the beam is prismatic).
2. The reinforcement is anchored properly into supports.
3. The capacity of the beam to resist shear forces does not control its strength.
4. Moments caused by self weight are negligible.

Estimate the magnitude of the maximum load P that the beam can resist.

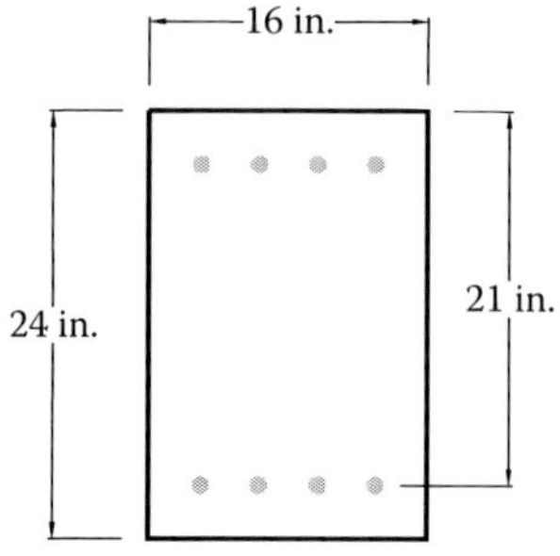

FIGURE 22.4 Cross section.

SOLUTION

This beam differs from the beams described previously in that the load is not symmetric with respect to a vertical line passing through mid-span. Nevertheless, the procedures described remain valid. The critical section for positive moment is not at mid-span, but at the point where the load P is applied. At the limit, and if there is sufficient rotational capacity, the moment at the section under the applied load and both end moments reach M_y (Figure 22.5):

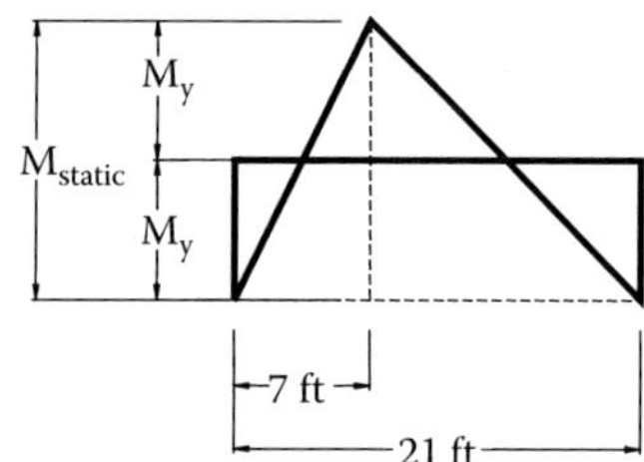

FIGURE 22.5 Final moment distribution.

$$M_{static\cdot} = 2M_y$$

The static moment is 2PL/(9), and therefore:

$$P = 9\ M_y/L$$

Remember that the yield moment and the nominal moment capacity are nearly equal. Before we estimate the moment capacity of the section we compute its reinforcement ratio:

$$\rho = A_s/(b \cdot d) = 4 \times 0.79 \text{ in.}^2/(16 \text{ in.} \times 21 \text{ in.}) = 0.94\%$$

Because the reinforcement ratio is neither excessive nor too small, we can estimate the moment capacity simply as

$$M_n = A_s \times f_y \times jd = 4 \times 0.79 \text{ in.}^2 \times 60 \text{ ksi} \times 0.9 \times 21 \text{ in.} \times 1 \text{ ft/12 in.} = {\sim}300 \text{ kip ft}$$

The load carrying capacity is therefore

$$P = {\sim}9 \times 300 \text{ kip ft/21ft} = {\sim}\ 129 \text{ kip}$$

BARE ESSENTIALS

- The limiting load capacity of a symmetrically loaded and reinforced concrete beam may be estimated closely by equating the static moment ($wL^2/8$ for uniform and $PL/4$ for concentrated load) to the absolute sum of a support yield moment and the yield moment at mid-span:

$$\frac{wL^2}{8} = M_{\text{yield}}^{\text{mid-span}} + M_{\text{yield}}^{\text{support}} \quad \text{for uniform load}$$

$$\frac{PL}{4} = M_{\text{yield}}^{\text{mid-span}} + M_{\text{yield}}^{\text{support}} \quad \text{for concentrated load at mid-span}$$

- For unsymmetrical loading or a reinforcement arrangement, an acceptable limiting load estimate can be obtained by using the appropriate expression for the static moment on the left-hand side of the equations above and using the mean value of the two end moments for the support moment.

EXERCISES

1. For the beam described in Example 22.1, draw an approximate moment diagram for the load that would cause first yielding at one of the critical sections. (See Example 21.1 to determine the initial magnitude of end moments.) Identify in the diagram the static moment.
2. Compute the load that would cause first yielding at one of the critical sections.
3. Determine the sequence in which critical sections of the beam described in Example 22.1 reach the yield moment.
4. Draw a sketch indicating where cracks would form as the beam is loaded to its maximum load.
5. What would the answer to Example 22.1 be if the load were applied at mid-span?

23 Combinations of Limiting Axial Force and Bending Moment for a Reinforced Concrete Section

Structural elements of reinforced concrete frames are likely to be subjected to combinations of axial force and bending moment. In this section, we shall develop a simple procedure to determine the limiting combinations of axial force and bending moment that a rectangular reinforced concrete section can resist. To simplify the process of calculations, we shall make three assumptions:

1. The moment acts about only one axis (Figure 23.1a).
2. The strain is distributed linearly over the section (Figure 23.1b), and the maximum compression strain is $\varepsilon_{cu} = 0.003$, the nominal limiting strain of concrete in compression.
3. Instead of assuming concrete stress is related to strain, we use the so-called rectangular stress block (Figure 23.1c). None of these assumptions limits the procedure that can be used for moments about two orthogonal axes, for non-rectangular sections, and for different stress–strain relationships, if desired.

Figure 23.2 shows three break points, corresponding to three states of stress, in a graph space defined by axial force and bending moment. For most sections used in practice, calculating three break points and connecting them with straight lines will suffice to obtain a satisfactory understanding of the limiting combinations of axial force and bending moment. The diagram that describes these combinations is called an interaction diagram.

We are already familiar with the determination of two of the points identified in Figure 23.2. Point 1 refers to limiting axial load. We learned how to determine that in Chapter 7 in relation to the axial load capacity of tied columns. Point 3 refers to the condition of bending moment. We studied that in Chapter 14. The only new item is point 2, which refers to the so-called balanced condition. Conventionally, the balanced condition is defined as the state with the limiting compressive strain at one edge of the section and the yield strain for the reinforcement at *the centroid of the tensile forces*. In sections with several layers of reinforcement, establishing the strain distribution that meets this definition requires iteration. In such cases, it is

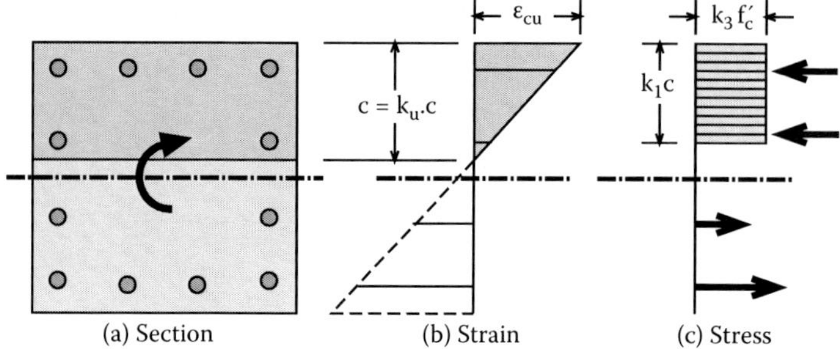

FIGURE 23.1 Reinforced concrete section subjected to axial force and moment.

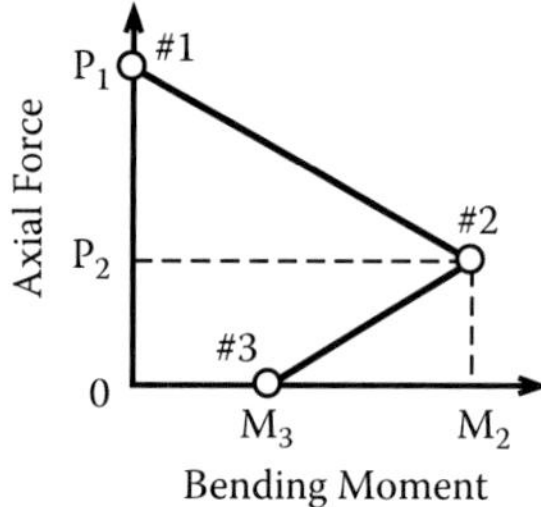

FIGURE 23.2 Interaction diagram.

more convenient to estimate point 2 as the point associated with a strain distribution having the limiting compressive strain at one edge of the section and the yield strain for the reinforcement at the *outermost layer of reinforcement in tension.*

Other points on the interaction diagram may be established by assuming a series of limiting strain conditions, determining the normal stresses on the section for that strain condition, and summing the stresses to determine the axial force and bending moment resistances.

We assume the reinforcement to respond linearly with increase in strain to reach a yield stress and then remain at that stress during further straining.

We start with point 1. As shown in Chapter 7, the nominal strength of the section in axial compression is

$$P_1 = \left(A_g - A_s\right) \times k_3 \cdot f'_c + A_s \cdot f_y \tag{23.1}$$

where P_1 is the strength of section in axial compression (point 1); $A_g = bh$, the gross cross-sectional area (Figure 23.3); $k_3 = 0.85$, the ratio of the strength of concrete in column to f'_c; f'_c is the compressive strength of 6 × 12 in. concrete cylinders; A_s is the total area of reinforcement; and f_y is the yield stress of reinforcement.

To determine point 2 we set the strain at $\varepsilon_{cu} = 0.003$ at the compressed edge and at

$$\varepsilon_y = f_y/E_s, \tag{23.2}$$

where ε_y is the nominal yield strain of the reinforcement and E_s Young's modulus for the reinforcement, at the outermost layer of reinforcement in tension (Figure 23.4b).

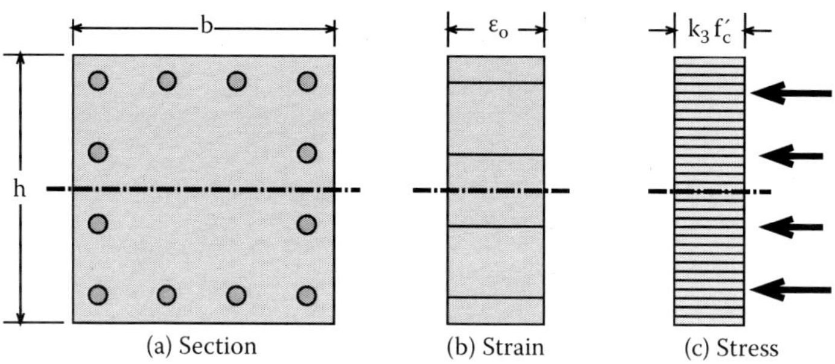

FIGURE 23.3 Unit strain and stresses at point 1.

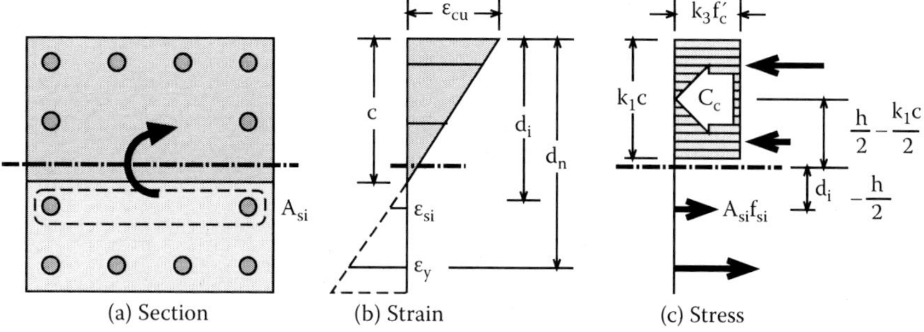

FIGURE 23.4 Unit strains and stresses at point 2.

For this condition, the depth to the neutral axis is

$$c = \frac{\varepsilon_{cu}}{\varepsilon_{cu} + \varepsilon_y} d_n \tag{23.3}$$

where d_n is the depth to the outermost layer of reinforcement in tension (Figure 23.4b).

The depth of the compression block is $k_1 c$ (Figure 23.4c).

For the assumed geometry and assumed strain condition, the force carried by the concrete is (Figure 23.4c)*

$$C_c = 0.85 \cdot f'_c \cdot b \cdot k_1 \cdot c \tag{23.4}$$

The forces carried by the steel depend on the magnitude of the strains in the steel. These strains may be computed as

$$\varepsilon_{si} = \frac{0.003}{c}(d_i - c) \tag{23.5}$$

where ε_{si} is the strain at the ith layer of reinforcement (Figure 23.4b) and d_i is the depth to the ith layer of reinforcement (Figure 23.4b).

---

* The area of the section in compression can be reduced to compensate for the area of reinforcement. Alternatively, the stress in the reinforcement in compression can be reduced by $0.85f'_c$ to achieve the same objective. This modification is correct but of trivial consequence.

For bars in compression Equation 23.5 yields negative numbers. The difference between the signs of strains computed for bars in tension and compression allows a calculation of the axial force at point 2 as follows:

$$P = C_c - \sum_i (f_{si} \cdot A_{si}) \tag{23.6}$$

where A_{si} is the area of the ith layer of reinforcement, C_c is the resultant of compressive stresses in the concrete, and f_{si} is the unit stress in the ith layer of reinforcement, given by

$$f_{si} = E_s \cdot \varepsilon_{si} \tag{23.7}$$

and

$$|f_{si}| \le f_y \tag{23.8}$$

The bending moment for point 2 is

$$M_2 = C_c \cdot \left(\frac{h}{2} - \frac{k_1 \cdot c}{2}\right) + \sum_i \left(f_{si} \cdot A_{si} \cdot \left(d_i - \frac{h}{2}\right)\right) \tag{23.9}$$

For point 3, the bending moment is determined with the limiting concrete strain at one edge of the section and with the neutral axis located so that there is no net axial force on the section.

The coordinates of points 1 and 2 are determined directly from the specified strain conditions. The easiest procedure to determine the moment for point 3 is by trial and error. The calculations described above are made for specific sections in the following example and Appendix A.

If a detailed definition of the interaction diagram is desired, calculations are made as illustrated in Appendix B.

EXAMPLE 23.1

Estimate the coordinates of point 2 for the section shown in Figure 23.5. Assume that the cross-sectional area of each bar is 1 in.^{2}, $f'_c = 4000$ psi, $f_y = 60$ ksi, and $E_s = 29000$ ksi.

SOLUTION

Yield strain

$$\varepsilon_y = \frac{f_y}{E_s}$$

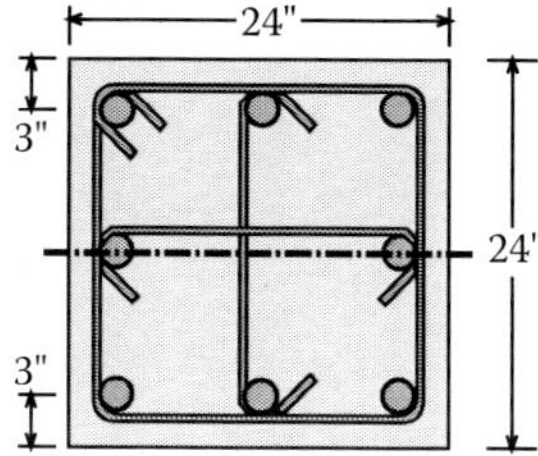

FIGURE 23.5 Column section.

Depth to neutral axis

$$c = \frac{0.003}{\varepsilon_y + 0.003} \cdot d_n \qquad c = 12.4 \text{ in.}$$

where $d_n = d_3 = 21$ in. is the distance from the outermost concrete layer in compression to the outermost reinforcing layer in tenstion.

Depth to other layers of reinforcement:

$$d_1 = 3 \text{ in.}$$

$$d_2 = 12 \text{ in.}$$

Steel layers are being numbered sequentially starting from the layer closest to the outermost concrete layer in compression.

Unit strains in reinforcement:

$$\varepsilon_{s_1} = \frac{0.003}{c} \cdot (d_1 - c) \qquad \varepsilon_{s_1} = -2.3 \times 10^{-3}$$

$$\varepsilon_{s_2} = \frac{0.003}{c} \cdot (d_2 - c) \qquad \varepsilon_{s_2} = -1 \times 10^{-4}$$

$$\varepsilon_{s_3} = \frac{0.003}{c} \cdot (d_3 - c) \qquad \varepsilon_{s_3} = 2.1 \times 10^{-3}$$

The negative signs indicate compression. The unit strain for layer 3 is equal to ε_y as intended.

Products of Young's modulus times unit strain:

$$E_s \cdot \varepsilon_{s_1} = -6.6 \times 10^4 \text{ psi}$$

$$E_s \cdot \varepsilon_{s_2} = -3 \times 10^3 \text{ psi}$$

$$E_s \cdot e_{s_3} = 6 \times 10^4 \text{ psi}$$

From inspection of these results we choose the following unit stresses in the reinforcement:

$$f_{s_1} = -f_y$$

$$f_{s_2} = E_s \cdot \varepsilon_{s2}$$

$$f_{s_3} = f_y$$

Observe that the signs match the signs of the strains and that the magnitude of no unit stress exceeds f_y.

Force resisted by the concrete:

$$C_c = 0.85 \cdot f'_c \cdot b \cdot k_1 \cdot c \qquad C_c = 862 \text{ kip}$$

Axial load:

$$P_2 = C_c - \sum_{i=1}^{3} \left(f_{s_i} \cdot A_{s_i}\right) \qquad \boxed{P_2 = 868 \text{ kip}}$$

Moment:

$$M_2 = C_c \cdot \left(\frac{h}{2} - \frac{k_1}{2} \cdot c\right) + \sum_{i=1}^{3} \left[f_{s_i} \cdot A_{s_i} \cdot \left(d_i - \frac{h}{2}\right)\right]$$

$$\boxed{M_2 = 753 \text{ kip} \cdot \text{ft}}$$

BARE ESSENTIALS

Combinations of axial load, P, and bending moment, M, that a section can resist can be estimated approximately by determining the P and M combinations for three simple conditions, (1) axial load, (2) balanced state, and (3) bending capacity, and connecting them with straight lines on a two-dimensional plot of P and M (Figures 23.2 to 23.4).

EXERCISES

1. Compute the moment–axial load coordinates of points 1 and 3 for the section described in Example 23.1.
2. Superimpose the coordinates of points 1, 2, and 3 on a complete interaction diagram generated by varying the depth to the neutral axis from 4 in. to 40 in.

24 Bond Properties of Plain Bars in Concrete

Bond refers to the interaction between concrete and reinforcement that is essential for the functioning of reinforced concrete as a composite. The physics of bond is not simple. It is quite similar to that of friction. It is convenient to explain friction as the product of a normal force and a coefficient. But it is not as simple, for example, to explain why the friction force is independent of the contact area. Bond has characteristics similar to those of friction. In this section, we discuss some of the factors that affect bond.

Currently, all but the smallest reinforcing bars have deformed surfaces to improve bond, *the transfer of force between concrete and reinforcement.* Nevertheless, we start by discussing the nature of bond in plain bars (in bars without intentional surface deformations) to find that the phenomena affecting bond strength in plain and deformed bars have many similarities. The differences are mainly in magnitude of unit bond strength, distribution of bond stresses along a bar, and the state of the surrounding concrete after a bar is pulled out of the concrete.

Figure 24.1 describes the surface of a plain bar (measurements made by G. Rehm, 1961). In microscopic scale, the surface of a plain bar making contact with concrete is not smooth. Cement paste is likely to penetrate into the microscopic crevices so that, initially, the force that resists relative movement (slip) between the bar and the surrounding concrete may be ascribed to two types of resistance: adhesion and shearing resistance of minute penetrations of concrete. Both types are related to tensile strength of concrete and to the force normal to the bar surface. The normal force is related to shrinkage or confinement by transverse reinforcement or applied force.

In 1913, Abrams published his monumental study of bond in reinforced concrete, in which he had studied the effects of (1) surface deformations of reinforcing bars, (2) embedment length, (3) size of specimen, (4) rust, (5) cross-sectional shape of reinforcement, (6) variation (taper) of bar along its length, (7) concrete age, (8) concrete mix proportions, (9) compressive strength of concrete, (10) curing of concrete, (11) end anchorage, (12) bar position (depth of concrete cast in one lift below bar), (13) autogeneous or self-healing of concrete, (14) repeated loading, (15) confinement by transverse reinforcement, (16) direction of casting of concrete, (17) shear span-to-depth ratio, (18) sustained load, and (19) lateral pressure during setting (Abrams, 1913). He left very little to be discovered by investigators who would follow him. Frustrated at not being able to distill a general principle (a feat that he was able to accomplish when he discovered the relationship of the water-to-cement ratio to compressive strength of concrete) from his 1610 bond tests, he ended his study with 59 finely crafted conclusions. Any engineer interested in bond should read Abrams's Bulletin 71, or at least his conclusions.

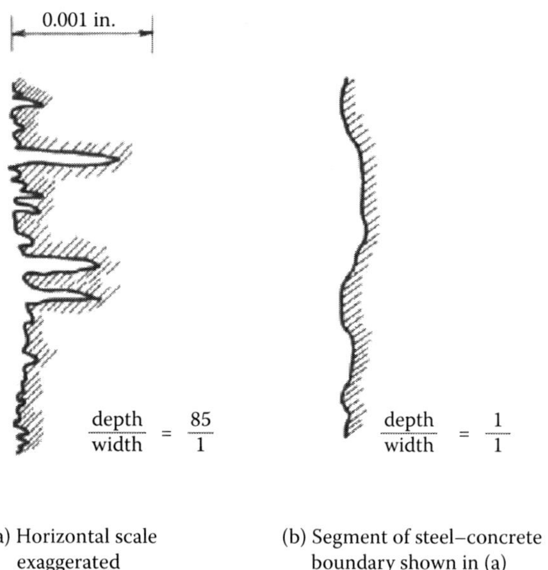

FIGURE 24.1 Surface of a plain bar in microscopic scale.

Bond stress is often considered in terms of the bond–slip relationship that appears to be akin to the stress–strain relationship for a material. But slip, reported in units of length, is not as convenient to generalize as unit strain, which has no units. Figure 24.2 shows a section of one of the standard test specimens that Abrams used. Bond stress is defined as the pull force divided by the surface of the bar in contact with concrete.

$$\mu = \frac{T}{\pi d_b L_d} \tag{24.1}$$

where T is the tensile force in reinforcement, d_b is the bar diameter, and L_d is the embedment length.

Slip could be measured at both the loaded and free ends (Figure 24.2). Abrams chose the slip at the free end as the defining measurement. Slip history at the free end would depend on the embedment length, the distance between the loaded and free ends. In order to make the definition of measured slip independent of embedment length, Abrams plotted the data from a series of pull-out tests as summarized below.

Take a bar with a diameter of, say, 1.25 in. and embed it in a 1500 psi concrete cylinder with a diameter of 8 in. Test it with an embedment length of 6 in. Plot the bond stresses measured at free-end slips of 0.0005, 0.001, 0.002, 0.005, 0.01, 0.02, 0.05, and 0.1 in. Repeat this procedure for bars having the same diameter and embedment lengths of 8, 12, 16, and 24 in. Project the plotted data back to the vertical axis to obtain a bond stress at an embedment length of zero as shown in Figure 24.3 for four values of slip. Plot the bond stress values so obtained against corresponding slip values, as shown in Figure 24.4.

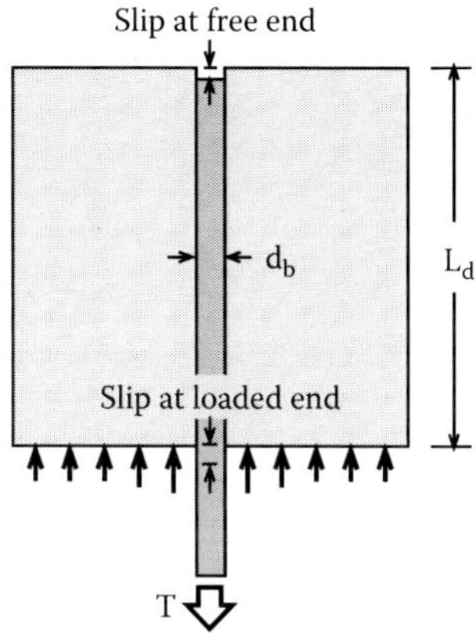

FIGURE 24.2 Vertical section of pull-out specimens tested by Abrams.

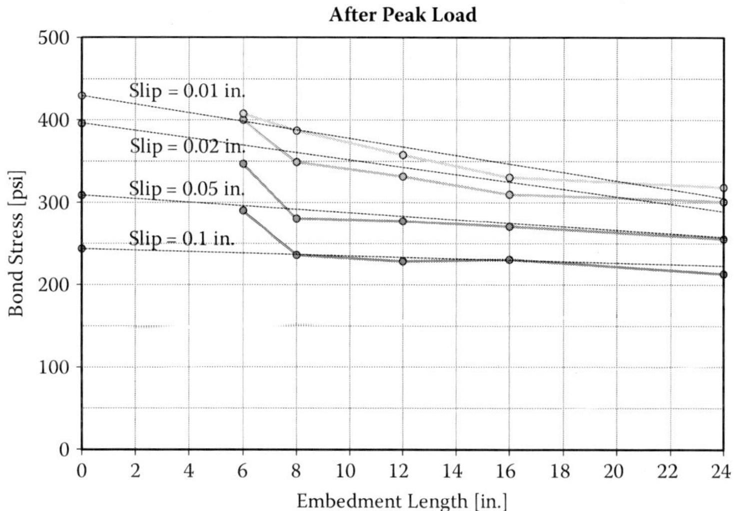

FIGURE 24.3 Bond stresses in plain #10 bars with variable embedment length.

Abrams confirmed the plot in Figure 24.4 for bars of different diameters and considered it to represent a general bond–slip curve for a plain bar. What is most important for the reader to note in his reasoning is that, over and above recognizing Abrams's creativity in overcoming the limits of his instruments, it is at best questionable to use bond–slip curves without knowing the details of how they were obtained.

Abrams concluded that at zero slip a bond stress of approximately 250 psi could be attributed to adhesion and static friction. In reference to Figure 24.1, we may consider the bond sources to be slightly different (the plain bar does have minute deformations and changes in diameter that make confinement important), but the approximate value that Abrams arrived at is still applicable. Abrams also noted that the bond stress increased as slip increased to approximately 0.01 in. He ascribed the

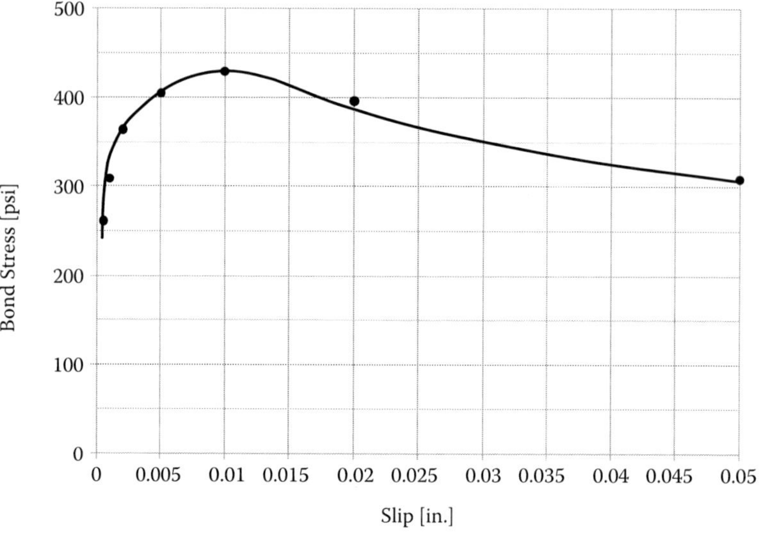

FIGURE 24.4 Bond–slip relationship proposed by Abrams for plain round bars.

increase to increasing normal force caused by unevenness in bar diameter along its length. Beyond a slip of approximately 0.01 in., bond resistance decreased and could be explained in terms of friction.

EXAMPLE 24.1

What is the required embedment length if a plain bar with a diameter of 1 in. is to develop a unit stress in tension of 40 ksi?

SOLUTION

The stress of f_s induces the tensile force of

$$T = \frac{\pi d_b^{\ 2}}{4} f_s$$

From Equation 24.1, based on the simplifying but incorrect assumption that the distribution bond stress is uniform along the length of embedment, we get

$$L_d = \frac{T}{\pi d_b \mu} = \frac{d_b}{4\mu} f_s$$

The actual peak bond stress along the bar is larger than 250 psi. Therefore, f_s can be larger than 40 ksi if

$$L_d \geq \frac{d_b}{4\mu} f_s = \frac{1 \text{ in.}}{4 \times 250 \text{ psi}} 40{,}000 \text{ psi} = 40 \text{ in.}$$

BARE ESSENTIALS

Bond stresses transfer forces from reinforcing bars to concrete and vice versa. In plain bars (bars without deformations), bond is related to adhesion, shearing of penetrations of concrete into minute crevices present on the surface of the bar, and friction.

Plain reinforcing bars embedded in normal weight aggregate concrete can develop a unit bond strength of approximately 250 psi at negligible slip. Initially, bond stress increases with slip by up to 50%. With additional slip, measured bond stress has been observed to decrease at a decreasing rate (Figure 24.4).

EXERCISES

1. Repeat Example 24.1 for a #4 bar.
2. What embedment lengths are required for #4 and #8 plain bars if they are to develop a tensile stress of 60 ksi?

25 Bond between Deformed Bars and Concrete

The relationship between bond stress and slip that Abrams (1913) inferred from his tests of deformed bars is shown in Figure 25.1. The shape of the curve indicates clearly that the distribution of the bond stress along the embedded length is certainly not likely to be anywhere near uniform. And yet, we shall find that in the design process we assume bond stress to be uniformly distributed along the length of a bar. Figure 25.2 contains three examples of current deformed bars.

The interaction of the deformations of the bar surface with the surrounding concrete may be visualized as illustrated in Figures 25.3 through 25.5. Figure 25.3 represents a section through an idealized deformed reinforcing bar embedded in concrete.

If a tensile force is applied at one end of the bar, the deformations tend to create a series of small and irregular wedges of concrete, as shown in Figure 25.4.

As the force on the bar is increased, the wedges push the surrounding concrete radially outward (Figure 25.5a). The radial expansion of the concrete leads to circumferential tensions or bursting forces (Figure 25.5b).

In a structural element, the process described above leads to splitting along the bar and may eventually cause a brittle failure.

With the caveat that any interaction of tensile stress and strength in concrete may lead to surprises, the following discussion provides a general understanding of phenomena related to circumferential tensions caused by bond around deformed bars.

It is plausible to think of the radial deformations caused by the deformed bar as having been caused by pressure within a tunnel. Let us consider the bottom tension flange of a reinforced concrete section with three deformed bars, as shown in Figure 25.6. Clear cover from the outside bars is A to the side and C to the bottom faces of the section. The clear distance between the bars is B.

In a linear homogeneous material, tensile stresses generated by the internal pressure would be high at the edge of the bar and would be expected to decay with distance from the bar. If length B is twice length A (side cover), the splitting crack may be expected to occur simultaneously on segments A and B or across the entire section. If length C (bottom cover) is shorter than length A (side cover), the splitting crack would be likely to appear on the bottom surface.

The splitting failure in Figure 25.7 emphasizes the question of bond stress distribution along a deformed bar. If one thinks of each deformation as developing a very large force at a very small slip, failure of the surrounding concrete is likely to be localized and proceed along the length of the bar progressively, as in the proverbial case of falling dominoes. In that extreme case, splitting could start at the end of the bar and continue rapidly along its length without developing a force sufficient to anchor the bar.

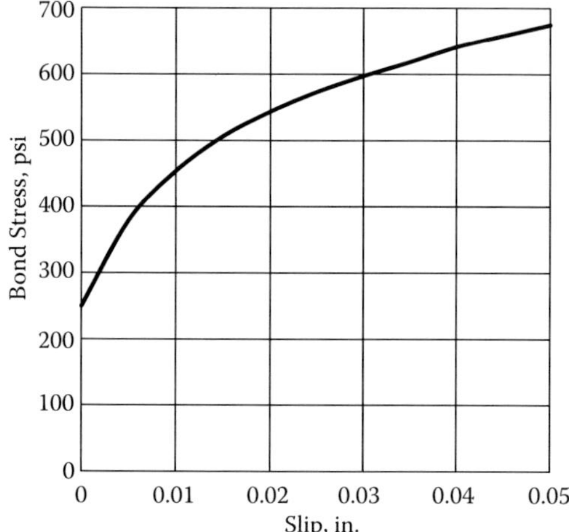

FIGURE 25.1 Bond–slip relationship for a deformed bar (Abrams, 1913).

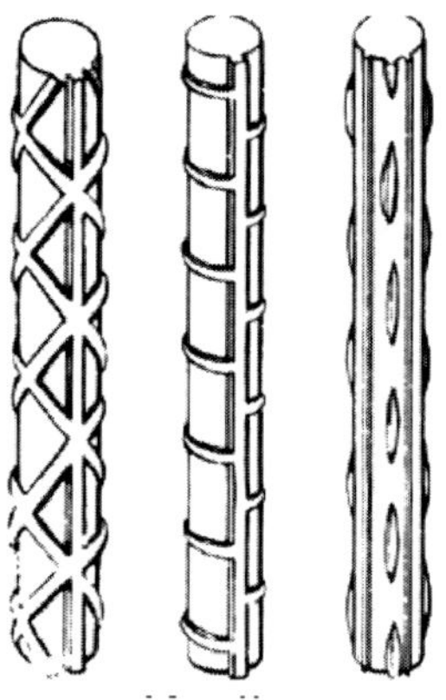

FIGURE 25.2 Representative deformed bars. (Note: The bearing area is usually smaller than 10% of the cross-sectional area of the bar.)

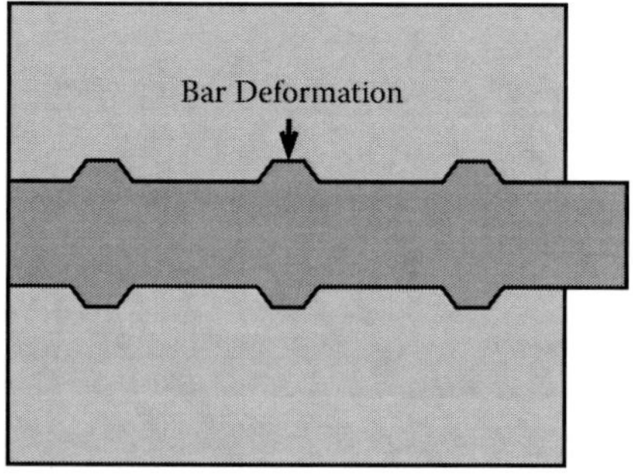

FIGURE 25.3 Idealized cross section showing a deformed bar embedded in concrete.

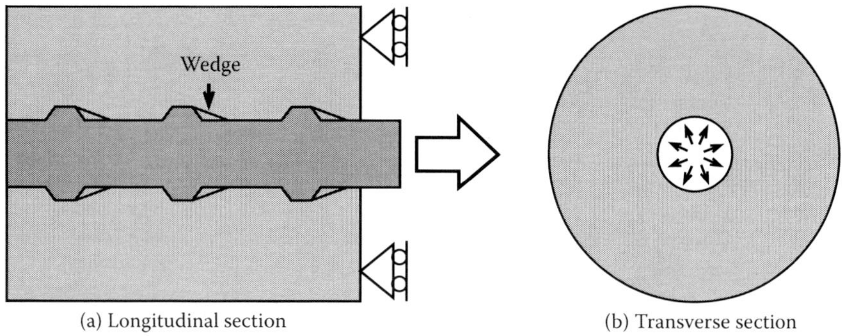

FIGURE 25.4 Bar deformations bear on the concrete to form wedges.

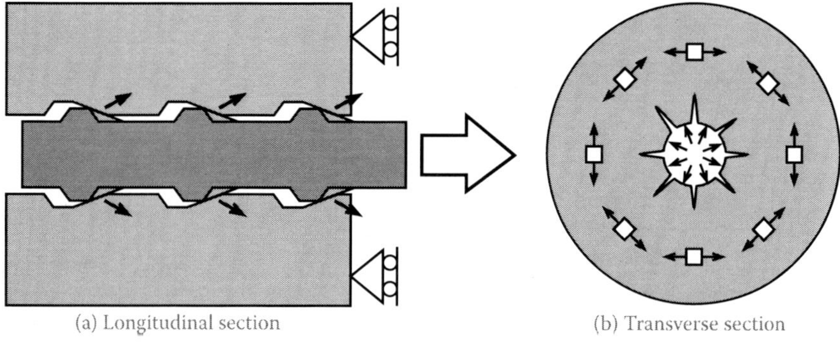

FIGURE 25.5 Relative movement of bar with respect to concrete, resulting in the wedges pushing the concrete out radially.

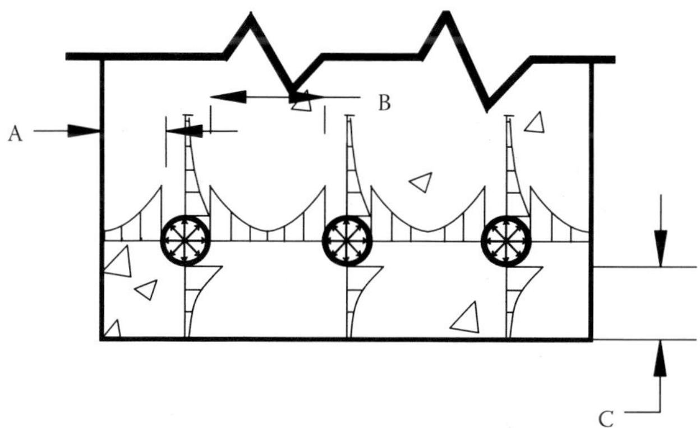

FIGURE 25.6 Idealized representation of tensile stresses around deformed bars.

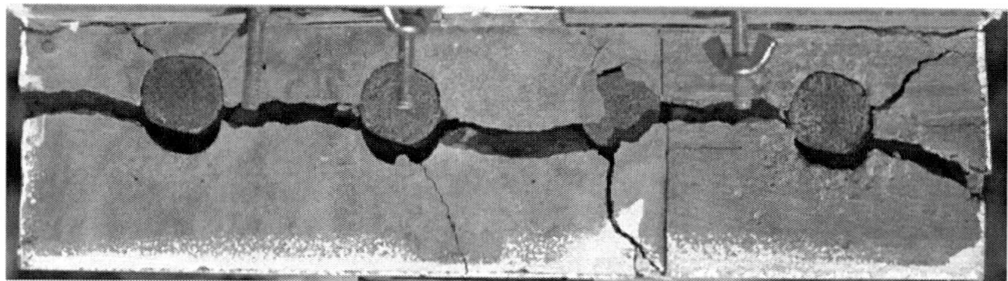

FIGURE 25.7 Splitting cracks caused by interaction between reinforcing bars and surrounding concrete.

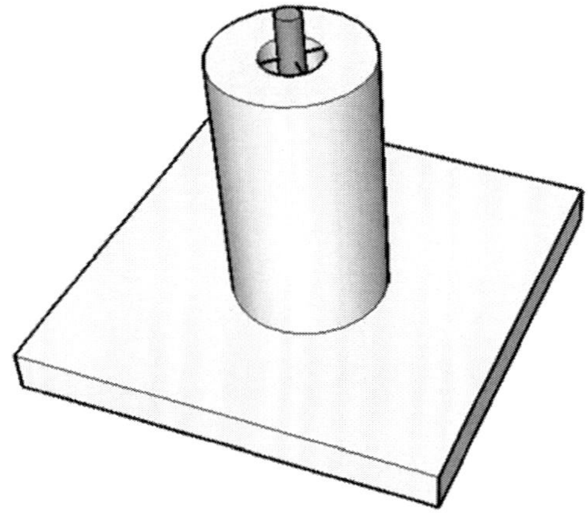

FIGURE 25.8 Metaphor for bond–stress distribution, a bar connected to a hollow concrete cylinder with a series of linearly elastic springs.

We are going to examine one dimension, and only one dimension, of the bond stress distribution problem by considering a metaphor: a reinforcing bar placed with a hollow cylinder of concrete and connected to it by springs that respond linearly (Figure 25.8). We assume the concrete to be rigid. A longitudinal section of the construct is shown in Figure 25.9a. It is assumed to be 10 units long. There are springs connecting the reinforcing bar to the concrete at each end of all units (at each node). The cylinder is attached to a fixed base.

If a tensile force is applied at the free end of the bar, the force will be transmitted to the cylinder at a rate depending on the longitudinal stiffness of the bar and the transverse stiffnesses of the connecting springs.

We can reason our way to one extreme condition. If the bar is rigid and the springs have finite stiffness, the problem is determinate. As the bar is pulled, the relative displacement (we can call it slip) will be the same at all points along the bar. Each spring will develop the same force. Projecting the metaphor to bond force, we can conclude that the distribution of the bond stress along the bar will be uniform.

If the ratio of the bar stiffness to the spring stiffness is finite, it is difficult to reach a confident answer for bond–stress distribution. But we can also hazard a guess about

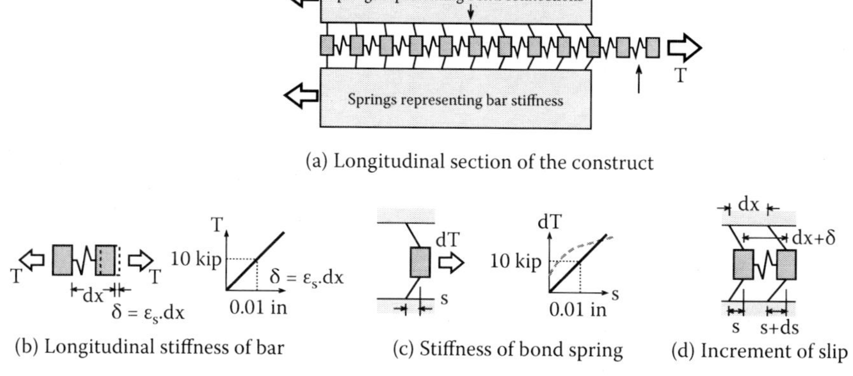

FIGURE 25.9 Metaphor to examine the bond–stress distribution problem.

another extreme condition. If the bar is very flexible (a bar made of rubber), the stress transfer is likely to be concentrated at the loaded end of the cylinder and is not likely to travel to the unloaded end. Note in Figure 25.9a that the springs near the loaded end are deformed more than those near the free end. In other words, slip is larger near the loaded end. The increment of slip ds in Figure 25.9d equals the elongation of the bar ($\delta = \varepsilon_s.dx$ in Figure 25.9b, where ε_s is the strain in the bar):

$$ds = \varepsilon_s dx \tag{25.1}$$

We may improve our guess by analysis of the metaphor in Figure 25.9a. We can calculate the force at each spring for given ratios of longitudinal bar stiffness to the sum of the transverse stiffnesses of the springs at each node*, κ. That should give us a measure of the bond stress distribution along the bar.

Figure 25.10 shows the distribution of the calculated spring forces along the embedded portion of the bar for three values of the relative stiffness factor, κ. The relative shapes of the curves indicate that unless the bar is almost rigid, the force tends to be transmitted to the concrete rapidly. That in turn suggests that the bond–stress around deformed bars is likely to be concentrated near the point of application of the tensile force until a splitting crack occurs. The initiation and the impact of the splitting crack would depend on confinement. If there is no confinement, the splitting crack may lead to failure.

We need to remember that we have assumed the connecting springs respond linearly, however high the relative displacement. If the springs yielded at a specified low displacement, uniform force distribution could be achieved. But even before

* Longitudinal bar stiffness is the force required for a unit longitudinal extension of the bar (slope of T-δ line in Figure 25.9b). The spring stiffness is the force required for a unit displacement of the spring in the direction of the bar (the slope of the solid dT-s line in Figure 25.9c). The actual response of the bond spring would be nonlinear, like the broken line of Figure 25.9c, which follows from Figure 25.1. Because bond strength depends on the tensile response of the concrete, it is important to note that the assumed connection would have a brittle fracture after limited straining.

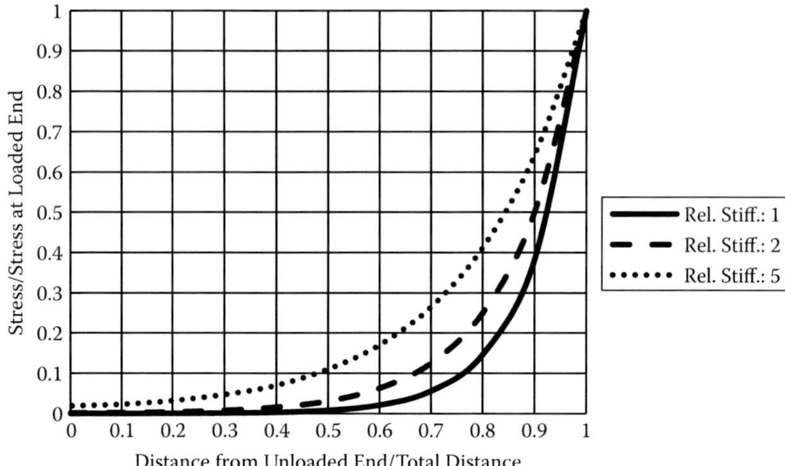

FIGURE 25.10 Force distributions calculated for $\kappa = 1$, 2, and 5.

studying experimental results, we understand that bond stress, intimately related to the tensile strength of concrete, might have a response resembling elastoplastic only under very special conditions of confinement. We admit that the metaphor we have used to understand the distribution of bond stress does not model the bond mechanism exactly, but it suggests that we must always treat the assumption of uniform bond stress with doubt. Furthermore, we should expect as much as 80% of the applied force to be transferred from the reinforcing bar to the concrete within 15 to 20 bar diameters of the loaded end.

A test carried out by R. M. Mains (1951) provides improved insight into the problem. The test specimen was a simple pull-out specimen, as shown in Figure 25.11. The embedded deformed reinforcing bar (#7 or 7/8 in. round) was equipped with strain gages along its length (Figure 25.12) installed inside the bar at a spacing of 2 in., starting at a distance of 1 in. from the free end of the bar.

Distributions of steel stress (based on measured strain) at two stages of the test are plotted in Figure 25.13. We note that a relative increase of 2.6 in the applied force (from 16 to 41 ksi) increased the apparent development length by 1.6 (from 21 – 11 = 10 in. to 21 – 5 = 16 in.), suggesting that the progressive failure we considered did not happen. The mean bond stress increased with increased force. We also note that the slope of the stress distribution curve is a measure of the bond stress, and that for the higher applied force, there is a reduction in the slope near the loaded end, implying the initiation of bond failure.

Unit bond stress μ is defined as the stress acting on the surface of the bar. It is obtained by dividing dT in Figure 25.9c by the area of the surface of the element:

$$\mu = \frac{dT}{\pi d_b \times dx}$$

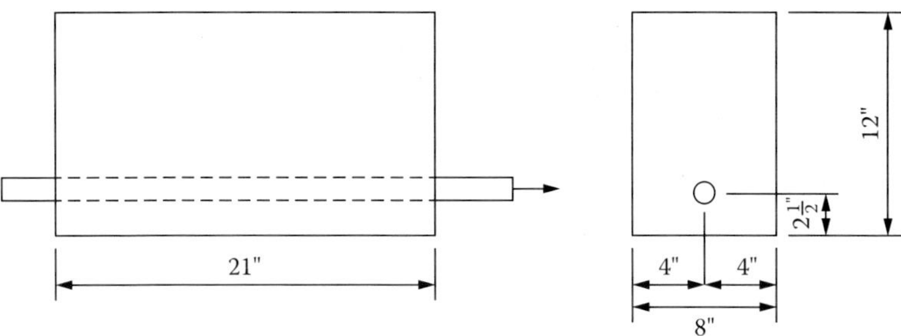

FIGURE 25.11 Pull-out specimen tested by Mains (1951).

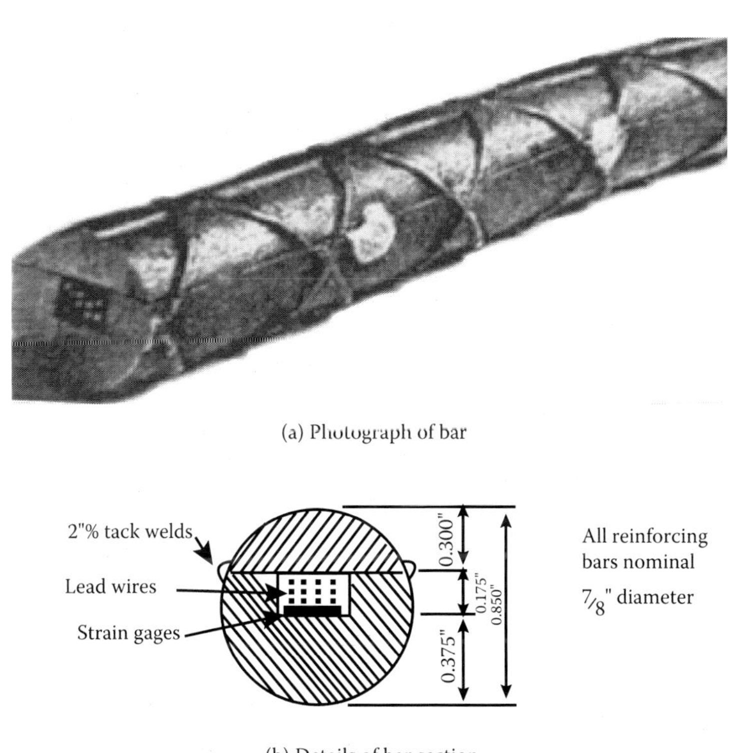

FIGURE 25.12 Reinforcing bar with strain gages along its length. (From Mains, R. M., American Concrete Institute, *Journal Proceedings*, 48, 11, 1951, 225–252, http://www.concrete.org/Publications/InternationalConcreteAbstractsPortal.aspx?m=details&i=11882.)

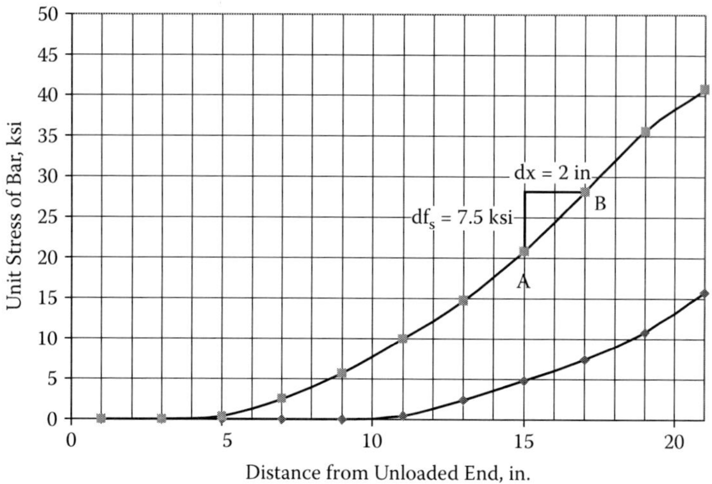

FIGURE 25.13 Distributions of steel stress at two stages of the test.

Because $dT = \frac{\pi d_b^2}{4} \times df_s$, we get

$$\mu = \frac{d_b}{4} \times \frac{df_s}{dx} \tag{25.2}$$

In the case of the change in stress between measurements A and B in Figure 25.13, we have

$$\mu = \frac{(7/8)\text{ in}}{4} \times \frac{7.5\text{ ksi}}{2\text{ in}} = 820\text{ psi}$$

We plot the bond stress distribution corresponding to the higher applied stress (41 ksi) in Figure 25.14.

Despite our reserve based on our knowledge that results of taking differences of measurements are quite sensitive to experimental error, we concede that the bond strength near the loaded end appears to have started decaying before the bond stresses at points closer to the unloaded end have reached their maxima. Also, as it follows from Abrams's insight, distribution of bond stress was not uniform along the deformed bar tested.

It is important to note that if we base the bond stress on the total length of embedment, the value for the case considered becomes 560 psi. It is evident from the data in Figure 25.13 that 560 psi is not the maximum bond stress. It is the mean bond strength based on a development length of 21 in. It should follow that if the development length had been shorter, the mean strength calculated at pull-out could have been determined to be higher. Refer to the work by R.W. Kluge and E. C. Tuma (1945) for information on the distribution of bond stresses along lap splices.

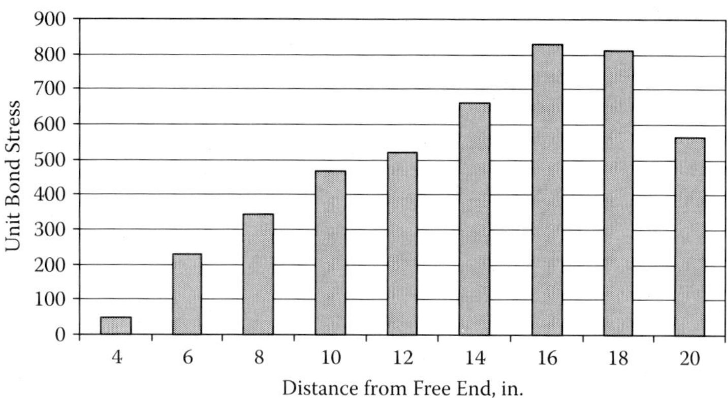

FIGURE 25.14 Bond stress distribution corresponding to the higher stress in Figue 25.13.

EXAMPLE 25.1

Compute the slip distribution along a deformed #7 reinforcing bar embedded in a 20 in. long prism (Figure 25.15a), assuming that the slip at the free end is zero and the bond stress is as follows:

Stage 1: $\mu = 0$ psi in $0 \leq x < 10$ in. and $\mu = 500$ psi in 10 in. $\leq x \leq 20$ in. (Figure 25.15b, left side).

Stage 2: $\mu = 500$ psi throughout the embedment length (Figure 25.15b, right side).

SOLUTION

We begin with Stage 1. Equation 25.2 leads to

$$df_s = \frac{4}{d_b}\mu dx$$

or

$$f_s = \int df_s = \frac{4}{d_b}\int \mu dx$$

where $\int \mu dx$ is the area of the shaded region in Figure 25.15b (left side). The steel stress at x = 20 in. is (Figure 25.15c, left side).

$$f_s = \frac{4}{d_b}\int \mu dx = \frac{4}{7/8 \text{ in.}} \times 500 \text{ psi} \times 10 \text{ in.} = 23 \text{ ksi}$$

The steel strain at x = 20 in. is (Figure 25.15d, left side)

$$\varepsilon_s = \frac{f_s}{E_s} = \frac{23 \text{ ksi}}{29 \times 10^3 \text{ksi}} = 0.0008$$

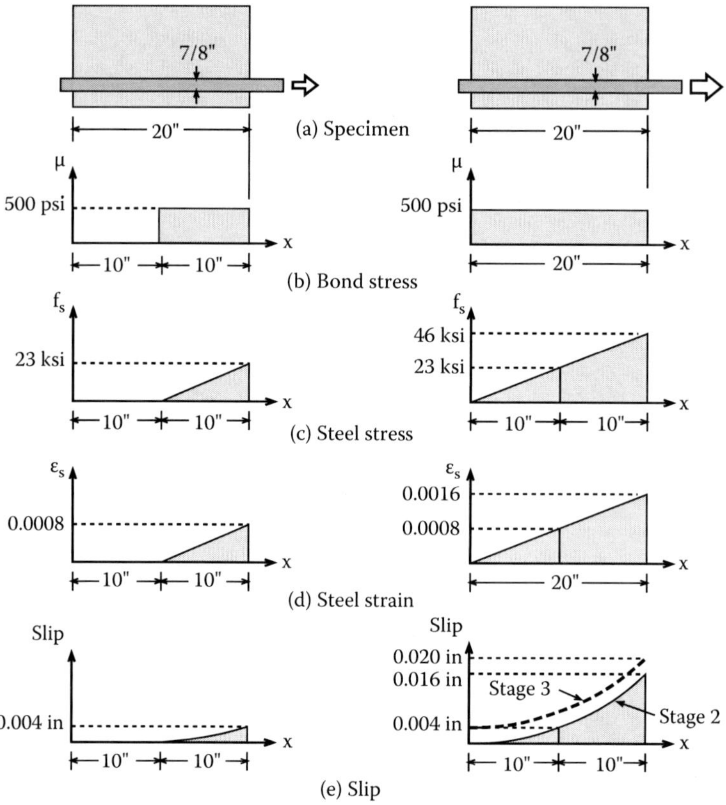

FIGURE 25.15 Stress and strain at Stage 1 (left side); stress and strain at Stages 2 and 3 (right side).

Equation 25.1 leads to $s = \int \varepsilon_s dx$, which is the area of shaded region in the diagram on the left side of Figure 25.15d. The estimated slip at x = 20 in. is (Figure 25.15e, left side)

$$s = \int \varepsilon_s dx = \frac{0.0008 \times 10\text{ in.}}{2} = 0.004\text{ in.}$$

Figure 25.16a illustrates Stage 1: Slip is zero between the free end and the middle ($0 \le x < 10$ in.), resulting in no bond stress and no steel strain; slip is still zero at x = 10 in., but a wedge is formed causing bond stress; and the slip at the loaded end is 0.004 in.

For Stage 2, we get the results on the right side of Figure 25.15c–e. Figure 25.16b illustrates Stage 2: Slip is zero at the free end (x = 0), but wedges form, and the slip at the loaded end is 0.016 in.

As we pull the bar further, slip occurs even at the free end, as shown in Figure 25.16c, where the slip at the free end is assumed to be 0.004 in. We call this hypothetical condition Stage 3. We assume that bond stress is still 500 psi, as shown in Figure 25.15b (right side), which causes the same steel strain as that

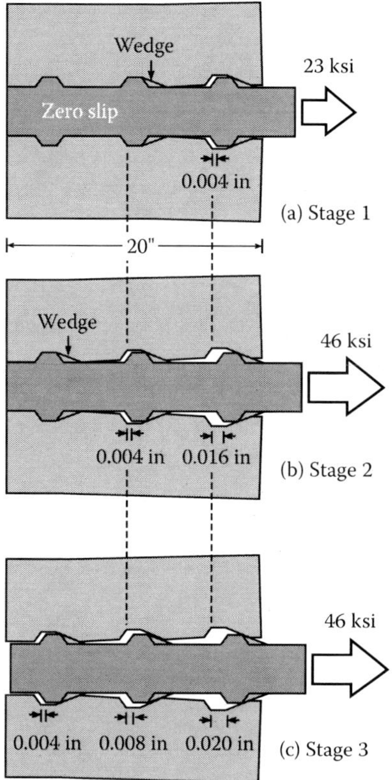

FIGURE 25.16 Wedge and slip.

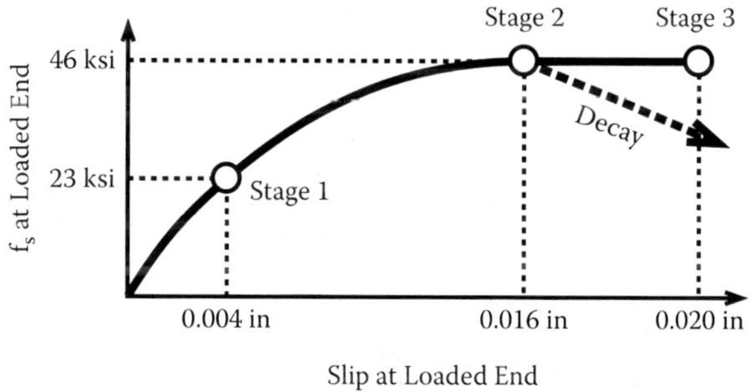

FIGURE 25.17 Relationship between steel stress and slip at loaded end.

of Stage 2 (Figure 25.15d, right side). Recalling ds/dx = ε_1 (Equation 25.1), we conclude that the slope of the slip distribution should be the same as that of Stage 2 and obtain the slip distribution indicated by the broken line in Figure 25.15e (right side). The slip at the loaded end is 0.004 + 0.016 = 0.020 in. The solid line in Figure 25.17 summarizes the relationship between the steel stress and the slip at the loaded end. In reality, however, the stress at the loaded end tends to decay, as indicated by the broken line, because of splitting failure (Figure 25.7).

BARE ESSENTIALS

1. Bond stress between deformed bars and concrete causes splitting cracks (Figure 25.7).
2. The distribution of bond stress along deformed bars is not uniform (Figure 25.14). It follows that longer development lengths are associated with smaller mean bond strength.

EXERCISES

1. Plot the bond stress distribution corresponding to the lower applied stress (16 ksi) in Figure 25.13.
2. Repeat Example 25.1 for a #10 bar.

26 Factors That Affect Bond

Mean bond strength data from 212 tests of black bottom bars (bars with untreated surfaces with concrete cover not smaller than one bar diameter and bar spacing not smaller than two bar diameters) of different diameters (from 3/8 to 1.7 in.) in normal weight aggregate concrete are plotted in Figure 26.1. The compressive strength of the concrete ranged from 2600 to 9900 psi.

The general trend of the plotted data confirms the earlier observation that the longer the embedment, the smaller the apparent mean bond strength. It is also noted that a mean bond strength of $6\sqrt{f'_c}$, with f'_c and the resulting stress in psi, represents a reasonable lower bound within the ranges of the variables included in the experiments and embedment or splice lengths not exceeding approximately forty bar diameters. All mathematical expressions in this chapter relate to stresses in psi units.

26.1 EFFECT OF COVER

We can infer that, given the criticality of the splitting stresses in the concrete surrounding the bar, an increase in side or bottom cover would tend to increase the bond strength. But given the scatter, it is not reasonable to consider these variables explicitly in design other than making certain that cover is not below one bar diameter and, preferably, two bar diameters. Building codes specify minimum values for cover and spacing that need to be followed in detailing in order to use the allowable bond strengths specified.

26.2 EFFECT OF TRANSVERSE REINFORCEMENT

Transverse reinforcement improves bond strength because it resists the bursting stresses.

Figure 26.2 shows bond strengths measured in 32 tests with varying amounts of transverse reinforcement (Sozen and Moehle, 1990). The horizontal axis represents values of the transverse reinforcement index defined as

$$\text{Index} = (A_{tr} \cdot f_y)/(N \cdot d_b \cdot s) \tag{26.1}$$

where A_{tr} is the total cross-sectional area of transverse reinforcement in each distance s and perpendicular to the bars being spliced or developed, f_y is the yield stress of transverse reinforcement, N is the number of longitudinal bars confined, d_b is the diameter of longitudinal bar, and s is the spacing of transverse reinforcement in the direction of the longitudinal bars.

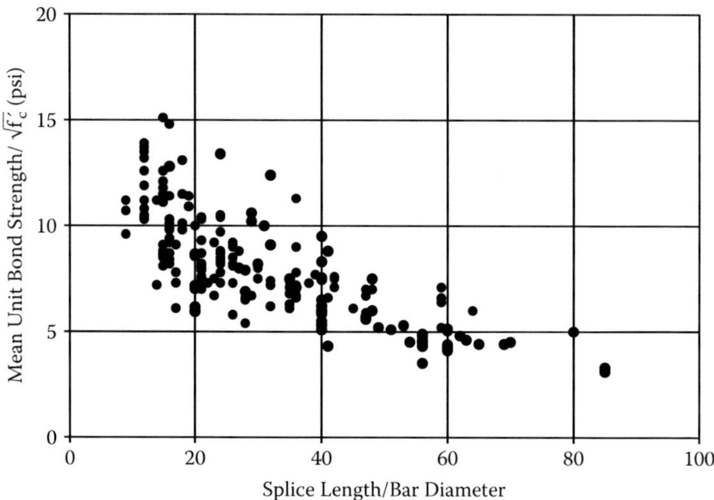

FIGURE 26.1 Data from tests of lap splices (with bottom-cast, unconfined, uncoated, deformed reinforcing bars—concrete strength not exceeding 10,000 psi, clear cover not smaller than one bar diameter, and clear bar spacing not smaller than two bar diameters).

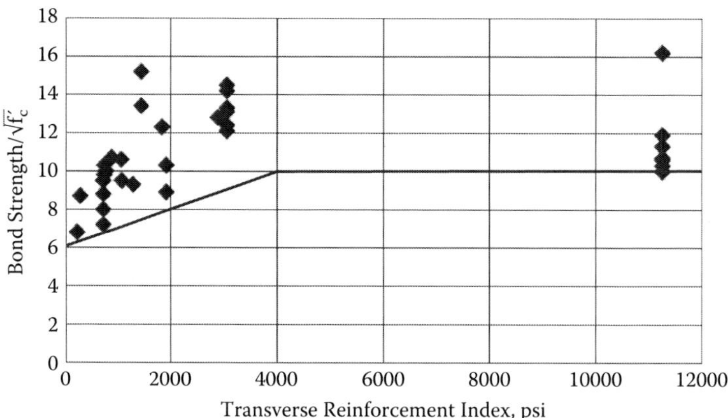

FIGURE 26.2 Bond strengths measured in 32 tests with varying amounts of transverse reinforcement.

For concrete cover not exceeding twice the bar diameter, it is reasonable to increase the bond strength to $10\sqrt{f_c'}$ linearly with the transverse reinforcement index up to an index value of 4000 psi.

$$\mu_a = \left(6 + \frac{\dfrac{A_{tr} \cdot f_y}{N \cdot d_b \cdot s}}{1000\ \text{psi}} \right) \cdot \sqrt{f_c'} \le 10\sqrt{f_c'} \tag{26.2}$$

where all stresses, including the result, are in psi. In design, this increase must not exceed the limit specified by the applicable building code.

26.3 DEPTH OF CONCRETE CAST BELOW REINFORCING BAR

The effect of having more than 12 in. of concrete cast in a single lift below the bar is shown in Figure 26.3 with the help of data from 36 tests (Sozen and Moehle, 1990). The bars represent the ratio of the strength measured for top bars to that of bottom bars. Considering that settlement of the concrete during setting will reduce the contact surface of the deformations with the concrete, a reduction is to be expected. Such a reduction was not measured in every case. The minimum ratio observed was 0.6.

26.4 EPOXY COATING

The effect of epoxy coating is likely to be quite sensitive to the thickness of the coating, the thickness of the bar deformations, and the quality and method of application of the epoxy. Tests reported by de Vries (1989) and Treece and Jirsa (1989) have registered ratios of bond strength of epoxied bars to that of black bars at values of approximately ½.

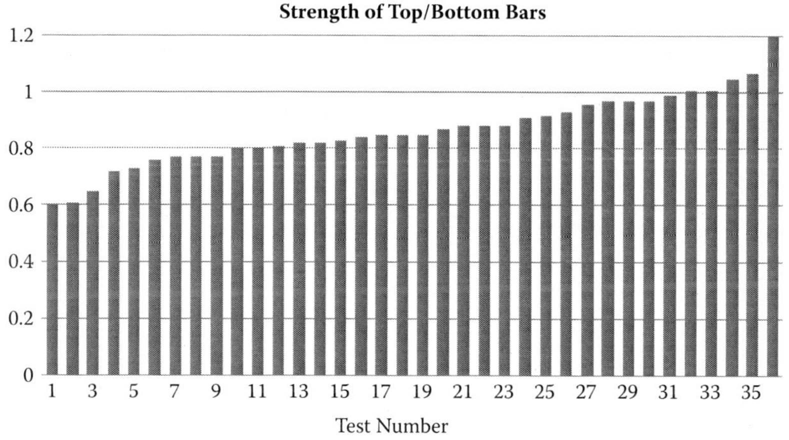

FIGURE 26.3 Ratio of bond strength of top bars to that of bottom bars.

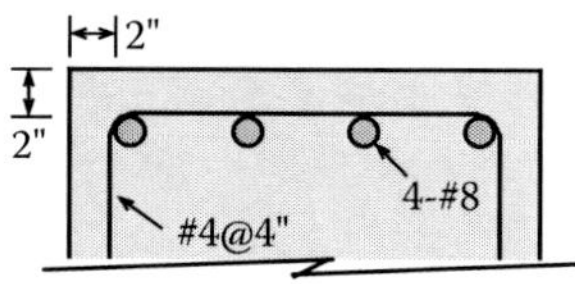

FIGURE 26.4 Top of beam section.

EXAMPLE 26.1

Estimate the maximum tensile stress that could be allowed in the longitudinal bars shown in Figure 26.4 if bond failure is to be prevented and the bars have a development length (length of bar embedded in anchorage or contiguous element) of L_d = 12 in. Assume f'_c = 5000 psi and f_y = 40 ksi. Assume that the bars do not have epoxy coating and the beam depth is 2 ft.

SOLUTION

The nominal cross-sectional area of a #4 bar is 0.2 in.^{2} (Chapter 4). The transverse reinforcement index is

$$\frac{A_{tr} \cdot f_y}{N \cdot d_b \cdot s} = \frac{\left(0.2 \text{ in.}^2 \cdot 2\right) \cdot \left(40000 \text{ psi}\right)}{(4) \cdot (1 \text{ in.}) \cdot (4 \text{ in.})} = 1000 \text{ psi}$$

From Figure 26.1 we estimate the mean bond strength as

$$\left(6 + \frac{\dfrac{A_{tr} \cdot f_y}{N \cdot d_b \cdot s}}{1000 \text{ psi}} \right) \cdot \sqrt{f'_c} = 7 \cdot \sqrt{f'_c} = 7 \cdot \sqrt{5000} \text{ psi} = 500 \text{ psi}$$

But because the beam depth is 2 ft, and therefore the depth of the concrete below the bars is more than 12 in., the mean bond strength should be set at 60% of 500 psi:

$$\mu_a = 0.6 \cdot 500 \text{ psi} = 300 \text{ psi}$$

The surface area of each longitudinal bar upon which this mean stress is assumed to act is

$$\pi \cdot d_b \cdot L_d = \pi \cdot 1 \text{ in.} \cdot 12 \text{ in.} = 38 \text{ in.}^2$$

The maximum allowable tensile force in each longitudinal bar is

$$T = \pi \cdot d_b \cdot L_d \cdot \mu = 38 \text{ in.}^2 \cdot 300 \text{ psi} = 11400 \text{ lbf} = 11.4 \text{ kip}$$

Because the nominal area of #8 bars is 0.79 in.[2], the corresponding allowable tensile stress is

$$f_s = \frac{11.4 \text{ kip}}{0.79 \text{ in.}^2} = 14.4 \text{ ksi}$$

The computed allowable bond strength of 14.4 ksi is much smaller than the yield stress (40 ksi), indicating that the bond may limit the strength of the beam. The development length should be increased.

BARE ESSENTIALS

In summary, we can conclude that for normal weight concrete the mean bond strength of black deformed reinforcing bars with proper cover (not smaller than one bar diameter) may be assumed to be $6\sqrt{f'_c}$ but not to exceed 600 psi unless directly relevant data are available. If the required development length exceeds 40 bar diameters, it is essential to seek relevant data.

If transverse reinforcement is used, the bond strength may be increased from $6\sqrt{f'_c}$ to $10\sqrt{f'_c}$ linearly with increases in transverse reinforcement index up to 4000 psi.

If more than 12 in. of concrete is cast below a bar in a single lift, the bond strength may be reduced to 60% of that for a bar with only a few inches of concrete cast below it.

Bond strength in lightweight aggregate concrete is quite sensitive to the properties of the aggregate. It can be comparable to or weaker than that for normal weight aggregate concrete of the same strength.

It is also important to remember that the basic value of $6\sqrt{f'_c}$ refers to a uniform distribution of bond stress along a bar in denial of what is observed in tests. It is a convenience to accommodate design. It does not represent the physics of the phenomenon, but it has served construction reasonably well so far. In all cases, the assumed bond strength must not exceed that allowed by the applicable code.

EXERCISE

Repeat Example 26.1 for the following cases:

	Longitudinal Bars		Transverse Reinforcement			Development
Case	Size	Number	Legs	Size	Spacing	Length
1	#8	4	2	#5	4	$20 \times d_b$
2	#8	4	3	#4	4	$20 \times d_b$
3	#9	3	3	#4	6	$40 \times d_b$

27 Design Examples for Bond

The inescapable conclusion from our review of experimental information on bond is that the distribution of bond stress along a reinforcing bar and its maximum value depend on complex interactions of many variables, such as bar profile, concrete strength, cover thickness, cracking, and confinement. Accuracy in predicting the magnitude and distribution of bond stress in an actual structure is not to be expected. The object of design calculations for bond is not prediction. It is to proportion a safe and serviceable structure. Procedures used for determining adequacy of bond are better understood and used in that light.

In the design environment, calculations are made for bond stress in relation to two different functions: (1) anchoring a bar so that the required strength is achieved at a specified section and (2) enabling the rate of stress transfer from a reinforcing bar to the surrounding concrete required by flexure.

The simple construct in Figure 27.1, a cantilever beam attached to a wall, illustrates the basic condition for anchorage. The reinforcing bar serving as the tensile element for the cantilever is anchored in the wall. The force in the bar at the edge of the wall must suffice to resist the moment produced by the loads resisted by the beam.

We start with the arbitrary assumption that the bond stress is distributed uniformly over the anchorage length (Figure 27.2). Considering a single bar and assuming that the allowable unit bond strength, μ_a, is known, the required anchorage length for a stress f_s to be developed in the bar is

$$L_d = \frac{\frac{\pi d_b^2}{4} f_s}{\mu \cdot \pi d_b} = \frac{d_b f_s}{4\mu_a} \tag{27.1}$$

where L_d is the required embedment length for anchoring the tensile force in bar AB, d_b is the bar diameter, and f_s is the unit tensile stress to be developed.

Having spent time to understand that the distribution of bond stress along an embedded bar, especially along a deformed bar, is far from being uniform, Equation 27.1 may appear to be in denial of the observed truth. And it is. At the same time, it is a very effective vehicle for understanding that not all assumptions of a design procedure need be in accordance with the facts.*

It is evident from Equation 27.1 that anchorage length should be determined by estimating a reasonable upper bound for the yield stress of the reinforcment and a reasonable lower bound for the unit bond strength assumed to be uniform. For a given bar

* In 1932 Hardy Cross wrote, "All analyses are based on some assumptions which are not quite in accordance with the facts. From this, however, it does not follow that the conclusions of the analysis are not very close to the fact."

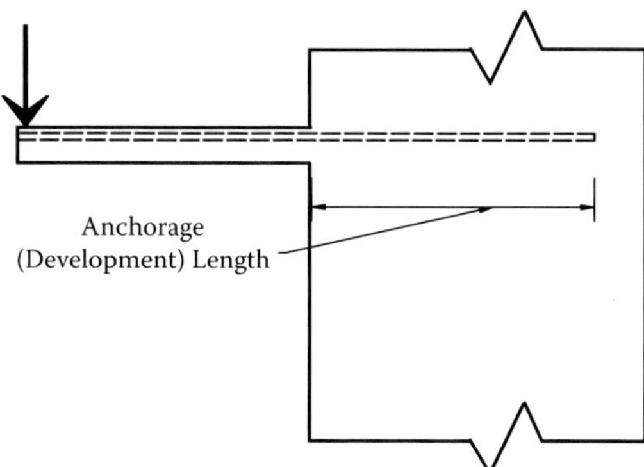

FIGURE 27.1 A cantilever beam attached to a wall.

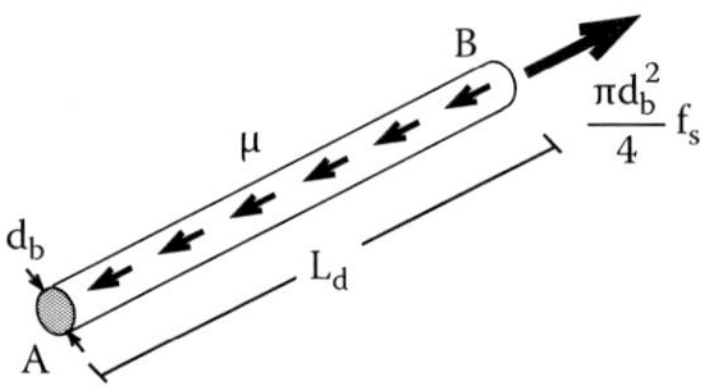

FIGURE 27.2 Bond stress distributed uniformly over anchorage length.

size, variations in diameter ought not to provide surprises. As discussed in Chapter 26, a reasonable lower bound to the allowable unit bond strength for deformed bottom black bars in normal weight aggregate concrete with proper cover and embedment lengths not exceeding approximately 40 bar diameters can be taken as

$$\mu_a = 6\sqrt{f_c'} \tag{27.2}$$

where f_c' is concrete cylinder strength in psi. We estimate the reasonable upper bound to the yield stress, f_s, to be $\frac{5}{4} f_y$.

From Equation 27.1:

$$\frac{L_d}{d_b} = \frac{5f_y}{96\sqrt{f_c'}} \tag{27.3}$$

If f_y is 60 ksi and f_c' is 4000 psi, the required anchorage length is 50 bar diameters. This value is neither general nor accurate, but it provides a basis for judging the adequacy of anchorage length. For 3000 psi concrete, the required L_d/d_b ratio approaches 60. It is also interesting to note that the force that is being anchored is the yield force. If it were desired to anchor the bar so that its strength is developed, the estimated length would increase by 50% or more. The required anchorage length may be reduced by confinement of the concrete surrounding the bar. In fact, from Equation 27.2 and Figure 26.1 it is apparent that our estimate of bond strength is not always conservative for long embedment lengths and using confinement would be prudent.

27.1 FLEXURAL BOND STRESS

Let us now consider the flexural bond stress requirement for the cantilever beam attached to the wall (Figure 27.3). Reinforcement is placed only near the top flange of the beam (Figure 27.4a). A concentrated load acts at the free end of the beam. If we ignore the self weight of the beam, the shear and moment diagrams are as shown in Figure 27.3.

To determine the bond stress required by flexure, we start with the assumption that the external moment is resisted by the internal tensile and compressive forces acting on the section at the edge of the wall. The internal forces should change along the span as the moment does, as long as the distance between the internal compressive and tensile forces, jd, does not change (Figure 27.4b).

$$dT = \frac{dM}{jd} \tag{27.4}$$

Recognizing that the change in moment is related directly to shear (Figure 27.3),

$$V = dM/dx \tag{27.5}$$

Substituting Equation 27.5 into Equation 27.4,

$$dT = \frac{V \cdot dx}{jd} \quad \text{or} \quad \frac{dT}{dx} = \frac{V}{jd} \tag{27.6}$$

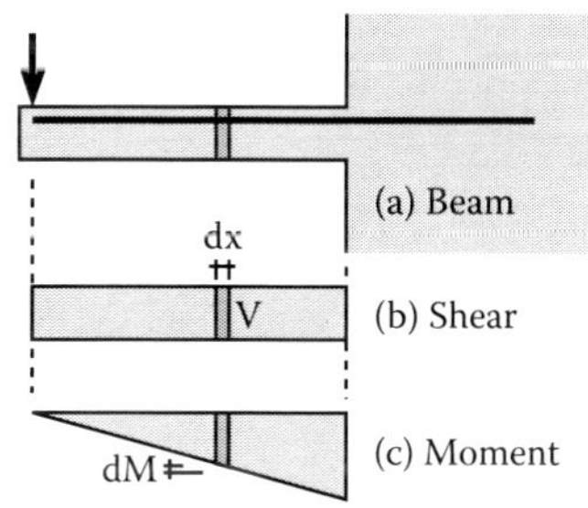

FIGURE 27.3 Cantilever beam subjected to a concentrated load.

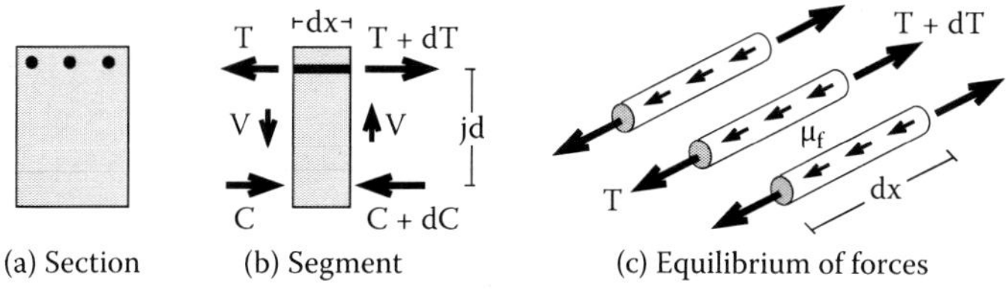

FIGURE 27.4 Equilibrium in infinitesimal segment.

Equilibrium of forces in Figure 27.4c leads to

$$dT = \mu_f \cdot N \cdot \pi d_b \cdot dx \tag{27.7}$$

where N is the number of bars. The flexural bond stress is

$$\mu_f = \frac{1}{N \cdot \pi d_b} \times \frac{dT}{dx} = \frac{V}{jd \cdot N \cdot \pi d_b} \tag{27.8}$$

The derivation that leads to Equation 27.8 ignores the effect of cracks along the length of the cantilever. Knowing the relationship between bond stress and slip for a deformed bar, we estimate that at the edges of each crack, the actual bond stress tends to reach its peak value. It is again denying reality to claim that the variation of flexural bond along the beam span is equal to that of the moment. But we make this assumption for design and understand that we are dealing with average phenomena and not with actual events.

Below we study three simple examples that cover the range of issues typically encountered in design.

EXAMPLE 27.1: CANTILEVER BEAM WITH A CONCENTRATED LOAD

A reinforced concrete cantilever, framing into a wall, has a span of 12 ft and cross-sectional dimensions 12 in. wide by 19 in. deep with an effective depth of 16 in. (Figure 27.5a and b). It is reinforced by two #9 Grade 60 reinforcing bars. Concrete strength for both the beam and the wall is 5000 psi. The beam is considered to be weightless. A concentrated vertical load is applied statically at the free end of the cantilever.

Conditions for bond may be considered independently of the magnitude of the applied load. In this simple case, we need not know what the design load is as long as we know the amount, quality, and arrangement of the reinforcement and the point of application of the load. The bond stress is checked assuming that the actual (and not nominal) yield stress of the reinforcement will be developed.

The two #9 bars are arranged in a straight line extending from the edge of the cantilever to 8 ft into the wall. The nominal yield stress of the bar is 60 ksi. But we

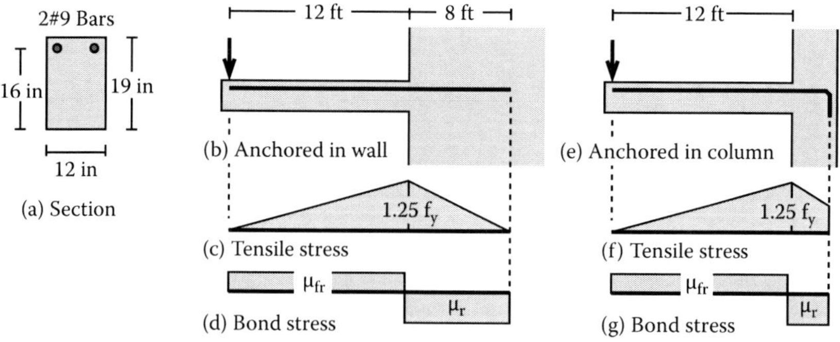

FIGURE 27.5 Reinforced concrete cantilever beam with a concentrated load.

believe that the best estimate for the bar yield stress is likely to be 60 ksi*(5/4) = 75 ksi. We use that value in evaluating the bond stresses.

The first calculation we make is a simple one. We do this to improve our sense of proportion. We note that the anchorage length of the bar (the length of bar embedded in the wall) is 8 ft. We check the ratio of the anchorage length to the bar diameter:

$$\frac{L_d}{d_b} = \frac{8\text{ ft}}{\frac{9}{8}\text{ in.}} = 85$$

Earlier we had estimated that for normal weight concrete an L_d/d_b ratio of 50 would be reasonable for deformed bottom bars. We project that to top bars, usually assumed to be bars with more than a foot of concrete cast below them, by estimating that the bond strength could decrease by 40%. Accordingly, the required L_d/d_b ratio may be close to 85. That result suggests that we need to look carefully into the question of required anchorage length and consider the use of transverse reinforcement.

The required anchorage length is computed assuming uniform distribution of bond stress along the anchorage length.

We had decided that for normal weight aggregate concrete a reasonable lower bound to bond strength is

$$\mu_a = 6\sqrt{f_c'}$$

For 5000 psi concrete, the expression would lead to approximately 420 psi. We also know that considering the 16 in. of concrete cast underneath the bar, we need to reduce that value by approximately 40%. The result is 250 psi.

We use the expression

$$L_d = \frac{d_b}{4\mu_a} \cdot \frac{5}{4} \cdot f_y = \frac{1.13\text{ in.}}{4 \times 250\text{ psi}} \cdot \frac{5}{4} 60\text{ ksi} = 85\text{ in.}$$

Comparing the calculated length with that available (8 ft = 96 in.), we decide that the anchorage is adequate. Nevertheless, every time we compute required development lengths exceeding 40 bar diameters we should consider confining the longitudinal reinforcement with hoops.

Next, we consider flexural bond. The resisting moment is estimated to be

$$M_{max} = A_s \times \frac{5}{4} f_y \times 0.9\text{ d} = \left(2\text{ in.}^2\right)\left(\frac{5}{4} \cdot 60\text{ ksi}\right)(0.9 \times 16\text{ in.}) = 180\text{ kip} \cdot \text{ft}$$

where A_s is the total area of tensile reinforcement and d is the effective depth of beam.

We determine the maximum shear, ignoring the effect of self weight. Shear at the face of the support is

$$V_{max} = \frac{M_{max}}{L} = \frac{180\text{ kip-ft}}{12\text{ ft}} = 15\text{ kip}$$

The resulting flexural bond stress, corrected to two significant figures, is

$$\mu_f = \frac{V_{max}}{jd \times N \times \pi \times d_b} = \frac{15 \text{ kip}}{0.9 \times (16 \text{ in.}) \times 2 \times \pi \times (1.13 \text{ in.})} = 150 \text{ psi}$$

which is satisfactory compared with the allowable value of 250 psi.

EXAMPLE 27.2: SIMPLY SUPPORTED BEAM WITH A CONCENTRATED LOAD

The beam considered (Figure 27.6a) spans 16 ft. Its cross-sectional dimensions (Figure 27.6b) are 14 in. (width) and 25 in. (depth). The effective depth is assumed to be 22 in. The compressive strength of the concrete is assumed to be 3000 psi. The beam is reinforced by three #9 bars of Grade 60.

Insofar as bond stress is concerned, there is a strong similarity between the cantilever we have considered in Example 27.1 and the simply supported beam we consider in this example, especially if the self weight of the beam is ignored. As long as the beam is loaded by a concentrated load at mid-span, each half of the beam about mid-span may be considered an inverted cantilever beam (Figure 27.6b to f). The main difference is that we need not be concerned with anchorage because ideally the forces on the bar at mid-span balance.

With that observation, all we need to do is check the flexural bond stress, which we know is given by the expression

$$\mu_f = \frac{V_{max}}{jd \times N \times \pi \times d_b}$$

As in the case of the cantilever beam, V_{max} is related to the maximum moment capacity of the section. First, we determine two dimensionless quantities, which will make us develop a sense of proportion essential for design. We determine the tensile reinforcement ratio.

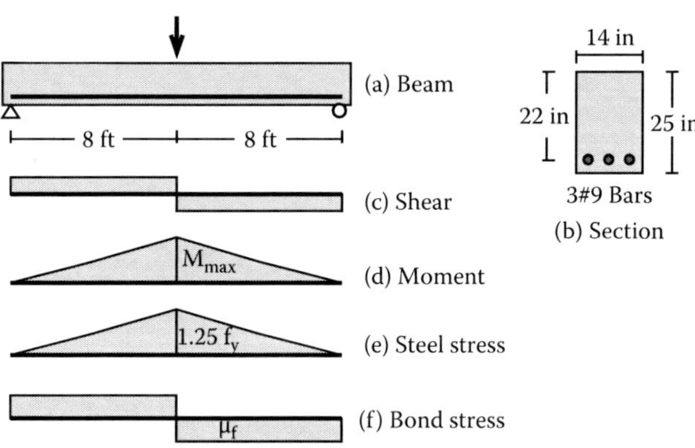

FIGURE 27.6 A beam with simple supports.

Section width, b = 14 in.
Effective depth of section, d = 22 in.
Cross-sectional area of tensile reinforcement, A_s = 2 in.[2]

$$\rho = \frac{A_s}{b \cdot d} = \frac{3 \text{ in.}^2}{(14 \text{ in.}) \times (22 \text{ in.})} = 1\%$$

Without making detailed calculations, we can estimate that a reinforced concrete beam with 3000 psi concrete and a tensile reinforcement ratio of 1% is likely to develop its flexural capacity.

We also note the L_d/d_b ratio for the bar:

$$\frac{L_d}{d_b} = \frac{8 \text{ ft} \times \dfrac{12 \text{ in.}}{1 \text{ ft}}}{1.13 \text{ in.}} = 85$$

However little experience we may have had examining L_d/d_b ratios, the L_d/d_b ratio of 85 determined above suggests that we should not have problems with bond stress.

We estimate the flexural strength of the section, assuming the yield stress may reach as high as (5/4) × 60 ksi = 75 ksi.

$$M_{max} = A_s \times \frac{5}{4} f_y \times 0.9 \text{ d} = \left(3 \text{ in.}^2\right)\left(\frac{5}{4} \cdot 60 \text{ ksi}\right)(0.9 \times 22 \text{ in.}) = 371 \text{ kip} \cdot \text{ft}$$

where A_s is the total area of tensile reinforcement, 3 in.^{2}; f_y is the nominal yield stress, 60 ksi; and d is the effective depth of beam, 22 in.

This results in a maximum shear force of

$$V_{max} = \frac{M_{max}}{8 \text{ ft}} = 46.4 \text{ kip}$$

Given V_{max} = 46.4 kip, the computed bond stress demand is

$$\mu_f = \frac{V_{max}}{jd \times N \times \pi \times d_b} = \frac{46.4 \text{ kip}}{0.9 \times (22 \text{ in.}) \times 3 \times \pi \times (1.13 \text{ in.})} = 220 \text{ psi}$$

The allowable limit for a deformed bottom bar in 3000 psi concrete with a cover of ~2 in. would be set at

$$\mu_a = 6\sqrt{f_c'} = 6\sqrt{3000} \text{ psi}$$

resulting in 330 psi. The calculated flexural bond stress demand is 2/3 of that allowable. There is no reason to expect problems related to bond in the simply supported beam considered.

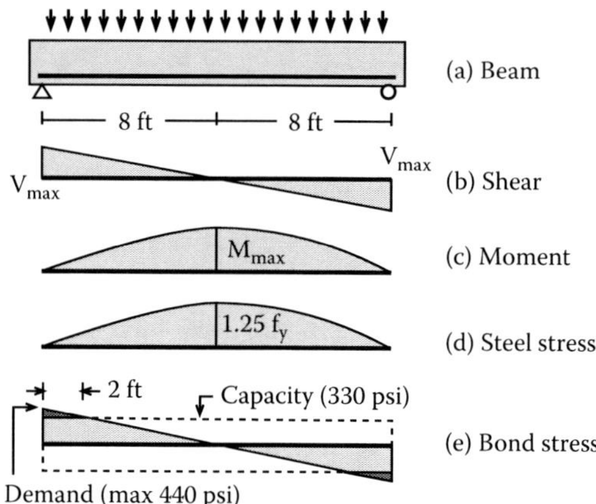

FIGURE 27.7 Simply supported beam with uniform load.

EXAMPLE 27.3: SIMPLY SUPPORTED BEAM WITH UNIFORM LOAD

In this example we consider the bond stress demand for the same beam but loaded uniformly as shown in Figure 27.7a. The moment capacity corresponding to a reinforcement stress of 75 ksi is the same, but the corresponding maximum shear changes to (Figure 27.7b and c)

$$V_{max} = \frac{w_{max}L}{2} = \frac{8M_{max}}{L^2}\frac{L}{2} = \frac{4M_{max}}{L}$$

And the maximum bond stress demand is

$$\mu_f = \frac{V_{max}}{jd \times N \times \pi \times d_b} = \frac{371\ \text{kip} \cdot \text{ft} \div (4\text{ft})}{(0.9 \times 22\ \text{in.}) \times 3 \times \pi \times (1.13\ \text{in.})} = 440\ \text{psi}$$

a value higher than the allowable unit bond strength of 330 psi (Figure 27.7e). There is a need to make changes in the properties of the beam. There are three practical options: (1) Use hoops to confine the bars at sections where the allowable bond strength is exceeded, (2) extend the beam over the supports to increase development length, and (3) use hooks at bar ends. We shall try each option.

1. Use hoops. We decide to add #5 Grade 60 hoops at a spacing of 4 in. We use the expression recommended for the effect of hoop reinforcement on bond strength (Figure 26.2)*:

$$\mu_a = \left(6 + \frac{A_{tr} \cdot f_y}{N \cdot d_b \cdot s} \cdot \frac{1}{\text{ksi}}\right) \cdot \sqrt{f'_c} \le 10 \cdot \sqrt{f'_c}$$

* During a professional application of the method, the relevant expression in the local code must be used.

where μ_a is the allowable bond strength of bars confined by hoops; A_{tr} is the cross-sectional area of both legs of one hoop, 0.6 in.^{2}; f_y is the yield stress of hoop bars, 60 ksi; N is the number of confined longitudinal bars (3); d_b is the diameter of longitudinal bars; s is the spacing of hoops (4 in.); and f'_c is the compressive strength of concrete.

The expression above leads to an allowable bond strength of 470 psi, which satisfies the demand of 440 psi. The hoops, which would also enhance the shear strength of the beam, are required in regions of the beam where the bond stress demand exceeds 330 psi (Figure 27.7e). The distance from the support to the point where the flexural bond is 330 psi is

$$8 \text{ ft} \times \left(1 - \frac{330}{420}\right) = 2 \text{ ft}$$

However, it would be appropriate to use the hoops over a quarter of the span, as shown in Figure 27.8.

2. Increase development length. The second option for meeting the bond strength requirement is to extend the beam at each support. The procedure is simple and straightforward, but it has a feature that we need to understand.

   The length to anchor the bar so it can develop (5/4) × 60 ksi at a bond strength of 330 psi is

$$L_d = \frac{d_b}{4\mu} \times \frac{5}{4} f_y = \frac{1.13 \text{ in.}}{4 \times 330 \text{ psi}} \times \frac{5}{4} \times 60 \text{ ksi} = 5 \text{ ft } 4 \text{ in.}$$

   and is shorter than half of the span (8 ft). But there is a catch. If we rely on anchorage bond and assume it to be uniformly distributed along the length of the bar, the available stress in the reinforcement at distance x from one end would be defined by Figure 27.9, or

$$f_s(x) - \frac{\pi d_b \mu_a x}{\frac{\pi d_b^2}{4}} = \frac{4\mu_a x}{d_b}$$

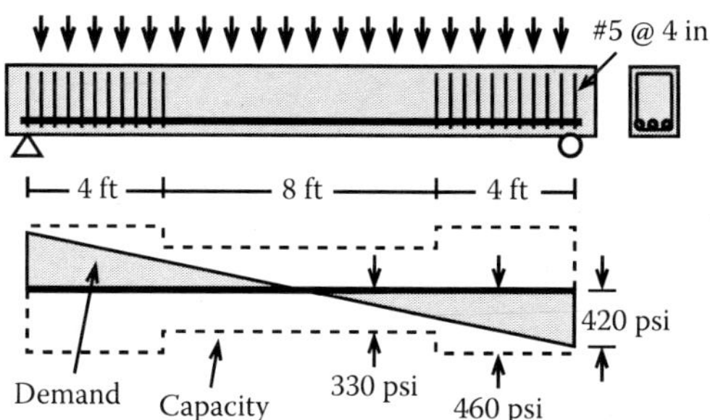

FIGURE 27.8 Use of hoops to increase bond strength.

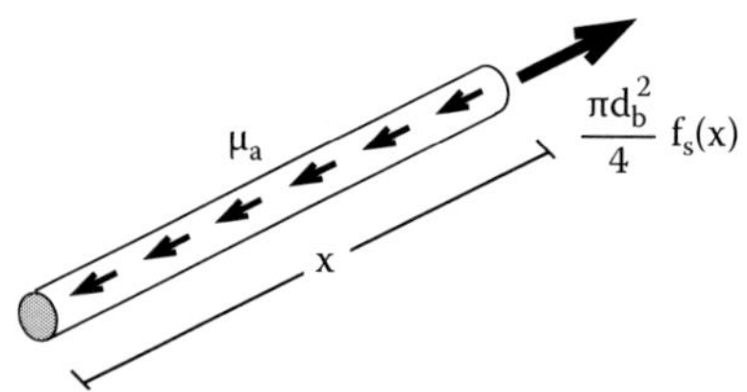

FIGURE 27.9 Available stress in the reinforcement at distance x.

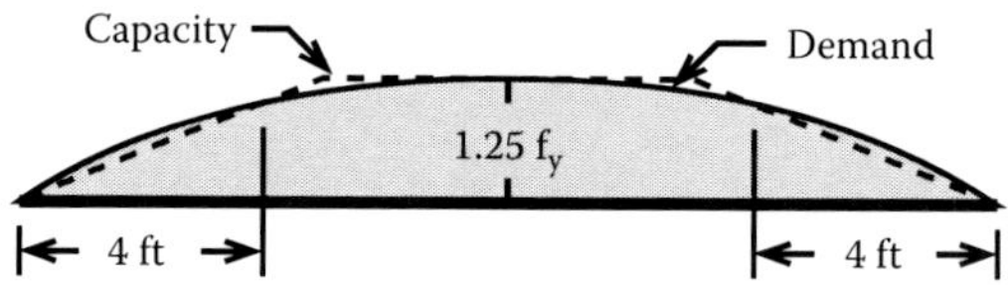

FIGURE 27.10 Demand and capacity of stress in reinforcement.

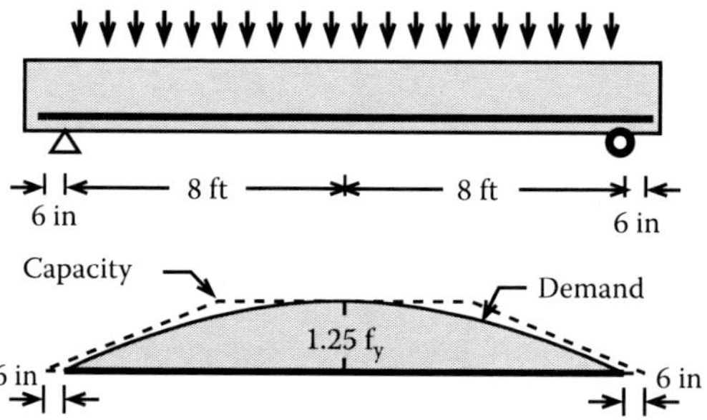

FIGURE 27.11 Extension of the bar at the reactions.

The broken line in Figure 27.10 shows the capacity determined by the previous equation. We note that up to approximately a distance of 4 ft from the support, the flexural stress demand exceeds the reinforcement stress accumulated through bond stress if no transverse reinforcement is used. The difference is not overwhelming, especially in view of the rather idealized assumption we have made about the distribution of bond stresses. Nevertheless, having committed to a set of assumptions, it is proper to be consistent in interpreting and applying the results. The relationship above indicates that we need to take measures to bring the solid curve below the broken curve at all distances from the reaction.

We decide to add 6 in. to the length of the bar at the reactions as shown in Figure 27.11. One may infer from the figure that extension could have been lower than 6 in. That is certainly correct within the set of assumptions we have used. But we also have to exercise a certain amount of judgment. In a problem involving bond, making small changes in development length may suggest that we do not understand the fundamental aspects of bond.

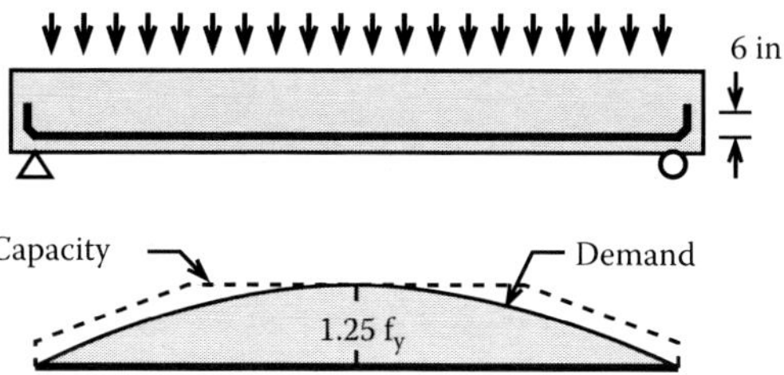

FIGURE 27.12 Hooked anchorage.

3. Use hooks. The third option is to provide a hook at each end of the bar, as shown in Figure 27.12. The available stress (the capacity) at the hook depends on various factors, including the concrete strength. A simplified approach to the determination of the required hook length is to assume that the bent part of the bar is equivalent to an extension of the anchorage length, as discussed above. A 6 in. vertical extension of the bar at each end would suffice to raise the resistance to that indicated by the broken lines in Figure 27.12.

   Of the three options discussed, adding hoops is likely to be the best solution to reduce the probability of bond failure.

BARE ESSENTIALS

- Anchorage bond stress is

$$\mu = \frac{\frac{5}{4} \cdot f_y \cdot d_b}{4 \cdot L_d}$$

- It is *not* acceptable to proportion a beam in which anchorage bond stress exceeds (on either side of each critical section) the allowable unit bond strength.

  Flexural bond stress is

$$\mu_f = \frac{1}{N \cdot \pi d_b} \times \frac{dT}{dx} = \frac{V}{jd \cdot N \cdot \pi d_b}$$

- It *is* acceptable to proportion a beam in which flexural bond stress exceeds the allowable unit bond strength as long as the reinforcement is extended past the inflection point so that the envelope representing moment (or flexural) capacity exceeds the envelope representing moment demands.

EXERCISE

For a 25 ft long beam in an exterior span of a continuous frame with a tributary width of 25 ft, select:

1. Cross-sectional dimensions.
2. Longitudinal reinforcement at critical sections (as integer numbers of standard bars).
3. Location of reinforcing bars at each critical section (draw and dimension cross sections to report your selection). Note: Select the clear spacing between bars to exceed two bar diameters and the clear cover to exceed one bar diamater.
4. Cutoff points for the longitudinal reinforcement selected (see Figure 3.3)
5. Draw envelopes for moment demand and moment capacity as shown in Figure 3.3. Note: Assume the slopes of the lines representing moment capacity near cutoff points are proportional to the allowable bond stress.
6. Anchorage length and hook dimensions for the bars anchored in the exterior column. Refer to the local building code to define the geometry of the hooks.

Assume

Yield stress of reinforcing bars = 60 ksi
Nominal concrete compressive strength = 5000 psi
Superimposed dead load = 120 psf
Live load = 50 psf
Strength reduction factor = 0.9
Controlling load combination: 1.2 (dead load) + 1.6 (live load)
Clear span = 25 ft
Column cross-sectional dimensions: 36 in. × 36 in.
Column vertical reinforcement: 18 #8 Grade 60 bars

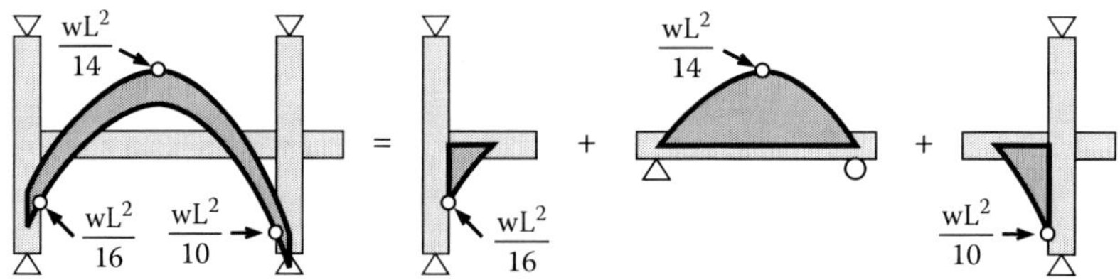

FIGURE 27.13 Beam decomposed into two cantilevers and a simple beam.

Hint: Treat the beam as two cantilevers supporting a simple beam (Figure 27.13). The cantilevers represent the segments resisting negative moments. The simple beam represents the segment resisting positive moments. Assume the distances from the column face to the points of inflection (where the bending moment is zero) are as follows:

Support	Segment	Distance to Inflection Point
Exterior	End (catilever)	$L_n/6$
Interior	End (cantilever)	$L_n/4$
Exterior	Middle (simple beam)	$L_n/10$
Interior	Middle (simple beam)	$L_n/10$

Note: L_n = clear span.

28 Control of Flexural Cracks

Widths of flexural cracks at service loads need to be considered in design (1) if the structural element is visible and (2) if it is exposed to a corrosive environment. One may get the impression from engineering literature that flexural cracking consideration in design requires calculation of crack widths to a precision of 0.001 in. We shall make no such pretense. Our goal is to provide a simple understanding of the dominant factors affecting flexural crack widths.

Tensile reinforcement cut from Grade 60 steel may sustain strains as high as 0.0015 at service load. The limiting tensile strain of concrete is unlikely to exceed 0.0002 and may be as low as 0.0001. Cracking of the concrete on the tension side of elements subjected to bending cannot be avoided, but crack widths may be controlled by proper proportioning.

To try and understand any poorly understood phenomenon, it is wise to look at its manifestations and see if results can be organized in terms of variables that affect them. Flexural cracking of reinforced concrete elements is certainly one of those poorly understood phenomena.

Figure 28.1 shows the cross-sectional properties of a test girder. Its elevation is shown in Figure 28.2, including the cracks in the central span between the applied concentrated loads. Thinking of the distribution of flexural strain over the depth of the girder, one would expect the crack width to increase at a constant rate from the neutral axis to the tension flange. Because of the stiffening effect of the reinforcement, the crack increases in width with distance from the neutral axis toward the tension flange, but is pinched at the level of the tensile reinforcement and starts expanding again as it approaches the extreme fiber in tension (Figure 28.3). In this chapter, the crack width we refer to is that at the level of the reinforcement.

The part of the span between the applied loads of the test girder in Figures 28.1 and 28.2 was subjected to essentially a constant moment at loads approaching and exceeding the service load.

The crack widths measured at the reinforcement level at both faces of the beam between the applied loads are shown in Figure 28.4. The widths shown were measured at reinforcement stresses of 31.2 and 36.4 ksi. Unquestionably, the most striking aspect of the crack width distribution is that the ratio of the maximum to the minimum measured crack width approached an order of magnitude at both levels of reinforcement stress.

Why should the scatter be so large? If we think of the cracking phenomenon as resulting from the transfer of stress from the reinforcement to the concrete surrounding it, the scatter appears quite plausible. The stress transfer must occur through a process akin to friction, and cracking must occur when sufficient stress has built up

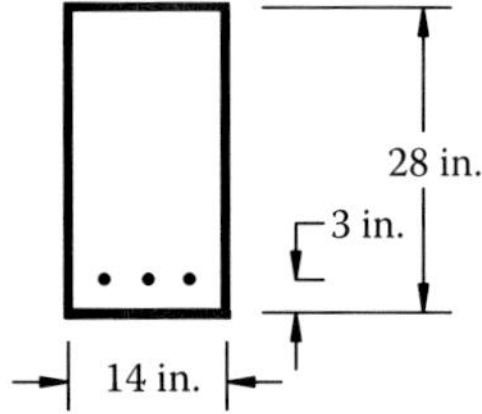

FIGURE 28.1 Section.

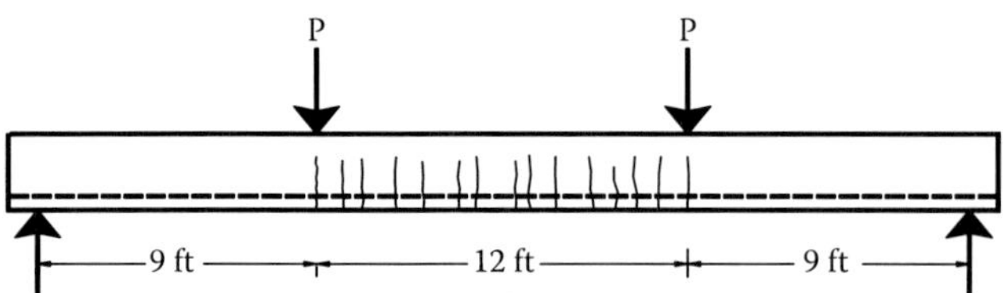

FIGURE 28.2 Cracks observed within the central span.

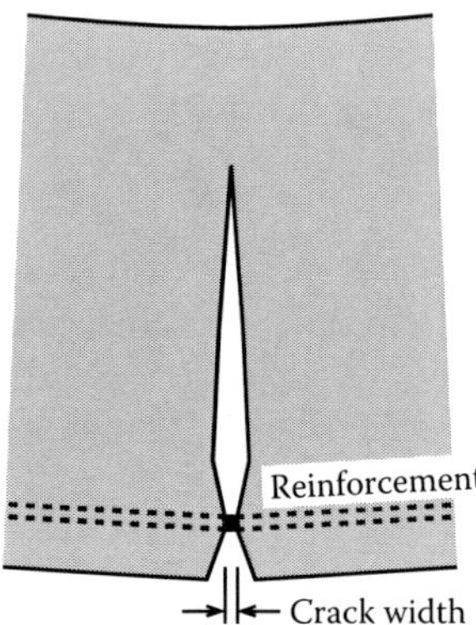

FIGURE 28.3 Detail of flexural crack.

in the concrete to reach its tensile strength. We have no reason to assume that either phenomenon is uniform along the beam. Crack spacing is bound to be random. If we think of the crack width as the accumulation of the strain over a distance equal to the crack spacing, it follows that it is unreasonable to accept all cracks as having the same width, even if the strain in the reinforcement is reasonably uniform.

Figure 28.5 shows the variation of crack width (at reinforcement level) with nominal reinforcement stress. The data indicate that there was a direct relationship between crack width and stress in the tensile reinforcement. We also note that for the test girder considered, the maximum crack width was approximately twice the mean crack width. It is also of interest to note that the reference crack widths, defined as the mean width plus two standard deviations, were comparable to the maximum crack widths up to a reinforcement stress of 40 ksi.

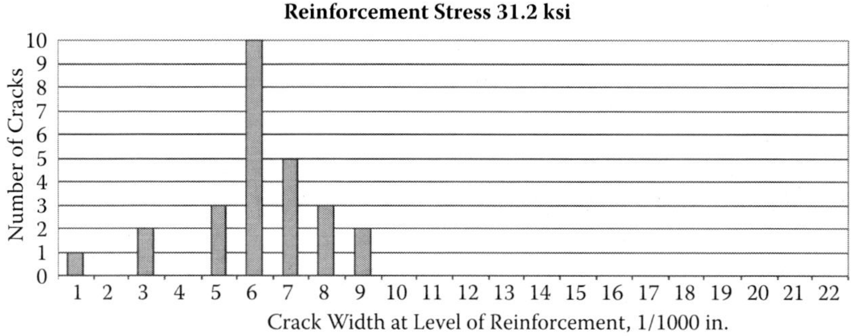

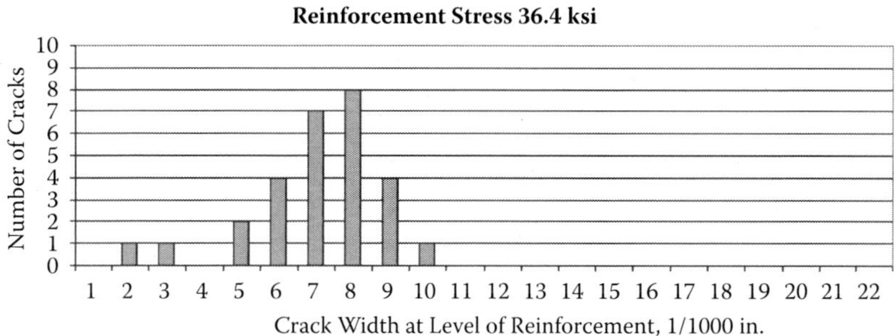

FIGURE 28.4 Crack widths measured at reinforcement level.

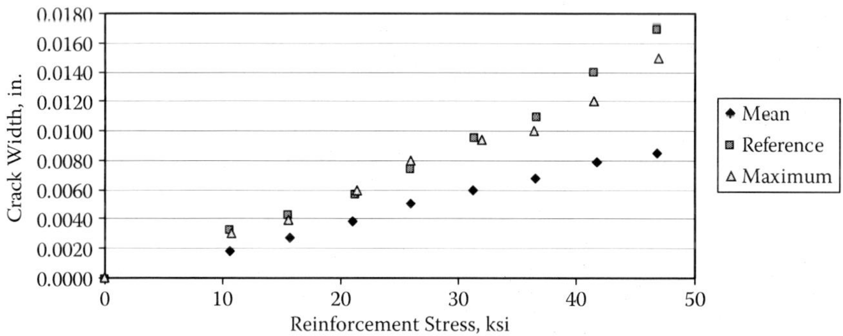

FIGURE 28.5 Variation of crack width with nominal reinforcement stress in the reinforcement.

To control cracking, we need to identify the dominant factors that drive the width of the crack. One of the factors is the tensile stress in the steel, as is evident from Figure 28.5. The width of the crack must be proportional to the extension of the reinforcement between two adjacent cracks. If we can identify the factors that influence crack spacing, we may be able to develop a procedure to control crack widths.

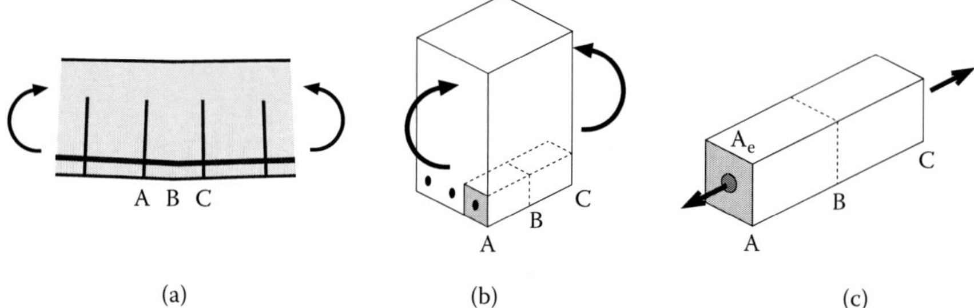

FIGURE 28.6 Search for the factors influencing crack spacing. (a) Flexural cracks. (b) Segment between two cracks. (c) A bar and concrete in tension.

To search for the factors influencing crack spacing, we consider a segment of a girder between two flexural cracks (Figure 28.6a and b) at A and C and ask whether another crack will form at mid-point B. We focus on one of the bars and the surrounding concrete defined in Figure 28.6c.

The tensile stress in the concrete at A, at one face of the existing crack, is zero. We reason that if the tensile stress at a vertical section crossing B can reach the effective tensile strength of the concrete, a crack may form at B (Figure 28.6a). The tensile force required for cracking at that section may be defined approximately as

$$TF_B = A_e \cdot f_t \tag{28.1}$$

where TF_B is the tensile force required to cause a flexural crack at B, A_e is the area on the vertical section at B over which the tensile stress exists, and f_t is the effective tensile strength of the concrete.

The reinforcing bar stress at A will be transferred to the concrete surrounding the bar along the length λ_{AB} by means of the bond stress μ (Figure 28.7b). Assuming the bond stress distribution to be uniform, we can express the total force transferred as

$$T_{tr} = \pi d_b \cdot \mu \cdot \lambda_{AB} \tag{28.2}$$

where T_{tr} is the tensile force transmitted from the reinforcement to the concrete per unit length, d_b is the diameter of bar(s), λ_{AB} is the distance from A to B, and μ is the bond stress on the surface of the bar acting along length λ_{AB}.

Ideally, a flexural crack will form at B if $TF_B = T_{tr}$, or

$$\lambda_{AB} = \frac{A_e f_t}{\pi d_b \mu} \tag{28.3}$$

The expression above does not justify our calculating the magnitude of λ_{AB}, but it does help organize the critical factors for determining the crack spacing. We go

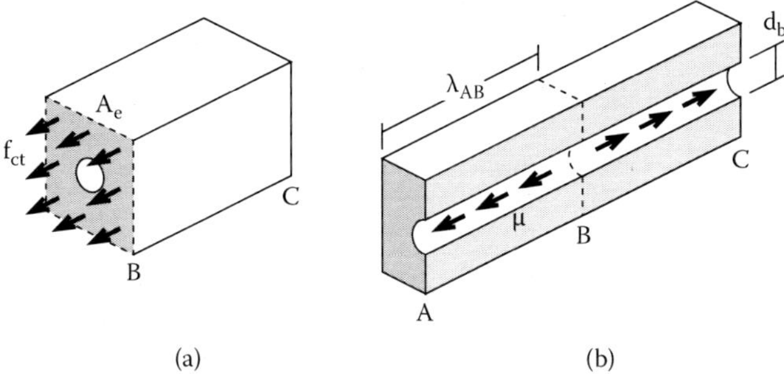

FIGURE 28.7 Tensile stress in concrete caused by bond stress. (a) Tensile stress in concrete. (b) Bond stress acting on concrete.

further. We assume that f_t and μ are equal. That assumption helps us reduce the expression to

$$\lambda_{AB} = \frac{A_e}{\pi \cdot d_b} \tag{28.4}$$

We do not know the area A_e, but we know that if we divide the area by d_b, we will obtain a length. What could that length be? We refer to an experimental study of cracking by Broms (1965) to infer A_e/d_b must be a function of the distance c_b from the extreme fiber in tension to the center of the reinforcing bar, provided the lateral spacing of the bars is approximately $2c_b$. Consequently, the mean crack spacing is also a function of c_b (Figure 28.8). Twice the concrete cover is a good starting estimate. For tensile reinforcement ratios below 1%, mean crack spacing has been observed to increase. This trend can be inferred from the reasoning that if the moment capacity of the section is equal to or smaller than the cracking moment, then ideally only one crack would develop in the beam.

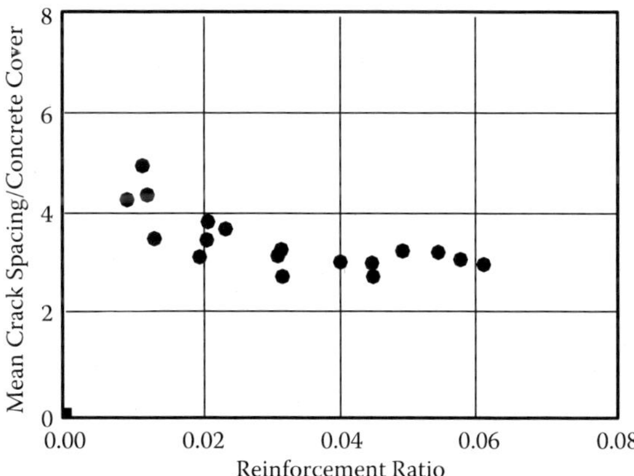

FIGURE 28.8 Mean crack spacing measured by Broms (1965).

We assume conservatively that the reinforcement strain is constant along the reinforcing bar. The mean crack width is then expressed as

$$w_m = 2c_b \cdot f_s/E_s \tag{28.5}$$

where c_b is the concrete cover, f_s is the stress in reinforcement at the cracked section, and E_s is Young's modulus for reinforcement.

The expression above identifies the dominant factors that control crack width and suggests the following design choices: (1) If the design criterion relates to visual appearance of the reinforced concrete member subjected to tensile stresses, the preferred action is to reduce the tensile stress (by adding reinforcement) to a minimum or to use the minimum cover required for other requirement reasons (such as fire protection) in the applicable building code. (2) If the design criterion is corrosion resistance, the reinforcement stress should be reduced by increasing the amount of reinforcement, but the cover, if at all possible, should be increased.

EXAMPLE 28.1

Estimate the mean and the maximum crack widths for the beam in Figure 28.2, assuming its tensile reinforcement is provided by three #10 bars and P = 30 kip.

SOLUTION

The bending moment is (Figures 28.2 and 28.9a)

$$M = 30 \times 9 = 270 \text{ kip-ft}$$

Assuming j = 0.9, the tensile force in the reinforcement is roughly (Figure 28.9b)

$$T = \frac{M}{jd} = \frac{270 \text{ kip-ft}}{0.9 \times (28 - 3) \text{ in.}} = \frac{270 \times 12}{0.9 \times (28 - 3)} \text{ kip} = 144 \text{ kip}$$

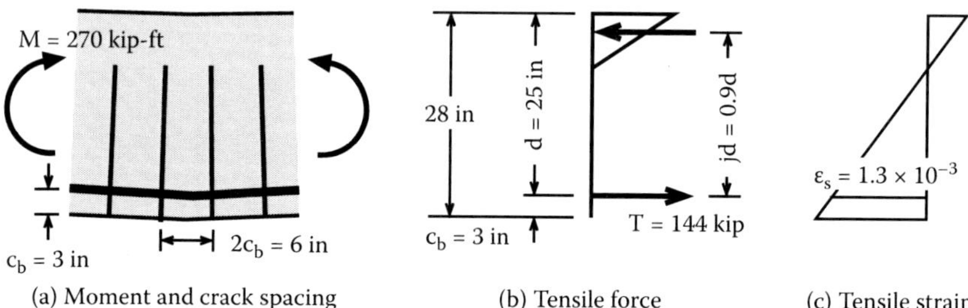

FIGURE 28.9 Crack spacing and tensile strain.

The stress in the reinforcement is

$$f_s = \frac{T}{A_s} = \frac{144 \text{ kip}}{3 \times 1.27 \text{ in.}^2} = 38 \text{ ksi}$$

The strain in the reinforcement is (Figure 28.9c)

$$\varepsilon_s = \frac{f_s}{E_s} = \frac{38 \text{ ksi}}{29 \times 10^3 \text{ ksi}} = 1.3 \times 10^{-3}$$

The mean crack spacing is (Figure 28.9a)

$$2 \times 3 \text{ in} = 6 \text{ in}$$

The mean crack width is

$$6 \text{ in.} \times 1.3 \times 10^{-3} = 8 \times 10^{-3} \text{ in.}$$

The maximum crack width is

$$8 \times 10^{-3} \text{ in.} \times 2 = 0.016 \text{ in.}$$

We should not be surprised if the actual maximum crack width might reach 0.03 in.

BARE ESSENTIALS

- The maximum crack width is not likely to exceed twice the mean crack width.
- The mean crack width may be estimated as a product of the mean crack spacing and the steel strain.
- The mean crack spacing may be assumed to be two to three times the concrete cover. Twice the concrete cover is a good, but not conservative, estimate for beams with a reinforcement ratio of more than 1%.

EXERCISES

1. Estimate the *mean* crack widths near critical sections for beam CD in Example 3.1. Assume $\rho = 0.01$ and $f'_c = 4$ ksi. Express your answer using an appropriate number of significant figures.
2. Estimate the maximum bar spacing that should be allowed in a slab if the *mean* crack width is to remain below 0.015 in. and the service level unit stress in the reinforcement is estimated to range from 30 to 40 ksi. Assume c_b in Equation 28.5 can be taken as half the bar spacing. Express your answer using an appropriate number of significant figures.
3. For Example 28.1, estimate the maximum load P that should be allowed if the *maximum* crack width is to remain smaller than 0.01 in. Express your answer using an appropriate number of significant figures.

29 Combined Bending and Shear

The topic that is often listed under the simple heading "Shear" in books and building codes is a design challenge in reinforced concrete created by the combination of shear and bending stresses. Consider the simply supported beam in Figure 29.1 with two concentrated loads placed at equal distances from the two reactions. The beam is homogeneous and weightless. Shear and moment diagrams for the loading assumed are included in Figure 29.1. We shall refer to the portions of the beam span subjected to shear as the shear spans without ignoring the fact that there is also moment in the shear spans. In the portion of the weightless beam between the applied loads, there is no shear and the moment is constant. We shall refer to this portion as the flexure span.

In the flexure span, there is no shear stress caused by the applied loads. The bending stresses are normal to the beam section. The principal stresses are the same as the bending stresses in magnitude and direction as shown at the extreme fiber in tension at mid-span in Figure 29.2.

In the shear spans, both bending and shear stresses exist (Figure 29.1b and c). At the neutral axis level, there is no bending stress and the principal tension stresses at that level are equal in magnitude to the shear stresses (Figure 29.1a), but are inclined at 45 deg to the longitudinal axis of the beam (Figure 29.2a).

To obtain an understanding of the general trend of the principal tension stresses, we can connect the directions of the principal tension stresses to obtain a principal tension stress trajectory, shown in Figure 29.2a by the broken line. If we repeat this for different values of the principal tension stresses, we obtain the trajectories in Figure 29.2b. Given those trajectories, we can reason that the cracks in the flexure span will be vertical, while those in the shear spans will tend to be inclined, as shown in Figure 29.3.

The principal tension stress trajectories provide an additional insight: Longitudinal reinforcement is assumed to provide resistance only in the direction of its axis. While it can compensate for the loss of resistance related to crack formation in the flexure span, it cannot do that efficiently in the shear span. Reinforcement is needed in another direction. The pioneers of reinforced concrete perceived this need and experimented with various arrangements of transverse reinforcement in addition to longitudinal reinforcement. Many configurations were tried (e.g., Figure 29.4) until construction requirements made vertical stirrups or hoops the most common type of transverse reinforcement. Several types of hoops and stirrups are shown in Figure 29.5. What differentiates the hoop from a stirrup is that the hoop bar is continuous on all four sides, and its ends are anchored within the core (the part enclosed by the hoop) of the section.

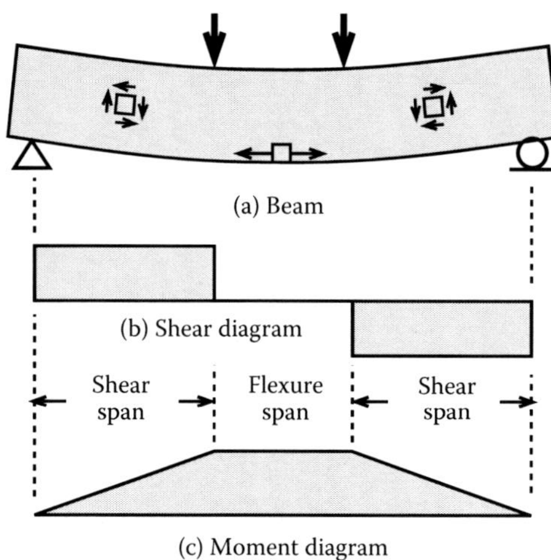

FIGURE 29.1 Stresses in simply supported beam.

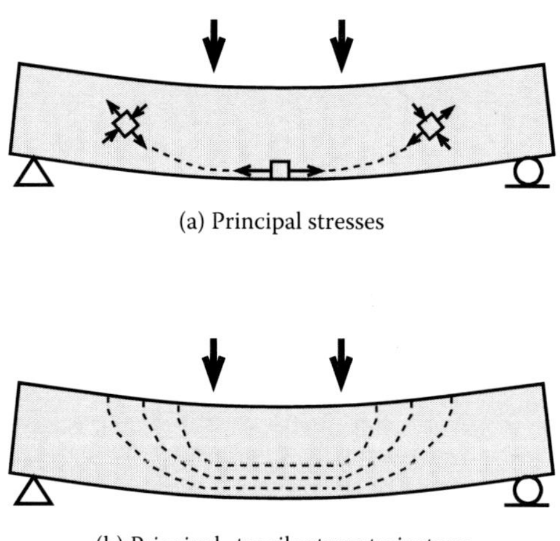

FIGURE 29.2 Principal stresses in simply supported beam.

Figure 29.5a shows an effective hoop. The longitudinal bars are contained by the hoop that has hook extensions into the confined core. The hoop in Figure 29.5b has its hook extensions in the shell of the section and becomes vulnerable if the shell concrete is lost. The use of hoops of the type shown in Figure 29.5a is required in elements resisting earthquake, blast, and impact effects.

The stirrup types shown in Figure 29.5b to e are used generally for construction in nonseismic regions. The hooks can be bent in (Figure 29.5c and d) or out (Figure 29.5e). A stirrup can be made to qualify as a hoop with an added crosstie. All building codes provide detailed specifications for the configuration of hoops and ties that must be followed to obtain approval for the structure.

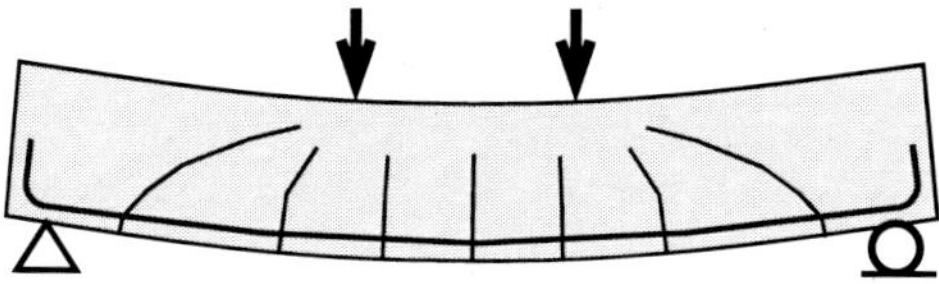

FIGURE 29.3 Cracks in simply supported beam.

FIGURE 29.4 Various arrangements of transverse and longitudinal reinforcement. (From Richart, F.E., An Investigation of Web Stresses in Reinforced Concrete Beams, *University of Illinois Bulletin* No. 166, University of Illinois at Urbana–Champaign, Urbana, Illinois, 1927.)

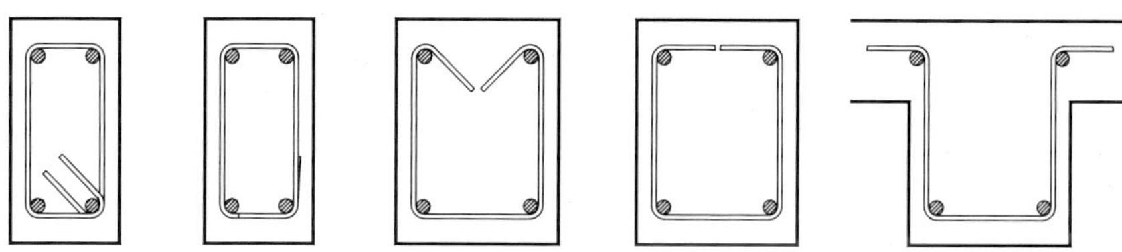

FIGURE 29.5 Several types of hoops and stirrups.

EXAMPLE 29.1

Compute the principal stresses and their directions at points A, B, and C in the cantilever beam shown in Figure 29.6. Assume that the beam is homogeneous, is uncracked, and responds linearly to stress. Neglect the self weight of the beam.

SOLUTION

1. The section modulus of the beam is

$$S = \frac{bh^2}{6} = \frac{12\text{ in.} \times (20\text{ in.})^2}{6} = 800\text{ in.}^3$$

The horizontal stress caused by the bending moment at A is

$$f = \frac{M}{S} = \frac{1000\text{ lbf} \times 36\text{ in.}}{800\text{ in.}^3} = 45\text{ psi}$$

At A, the stress caused by the shear force is zero. The vertical stress also is zero. Figure 29.7a shows the stresses acting on an infinitesimal square at A. Figure 29.7b shows the corresponding Mohr's circle. Point P in Figure 29.7b represents the stresses on the vertical faces (P) in Figure 29.7a. The x-coordinate of point P in Figure 29.7b is positive because the normal stresses on faces P are tensile. The y-coordinate of point P is zero because

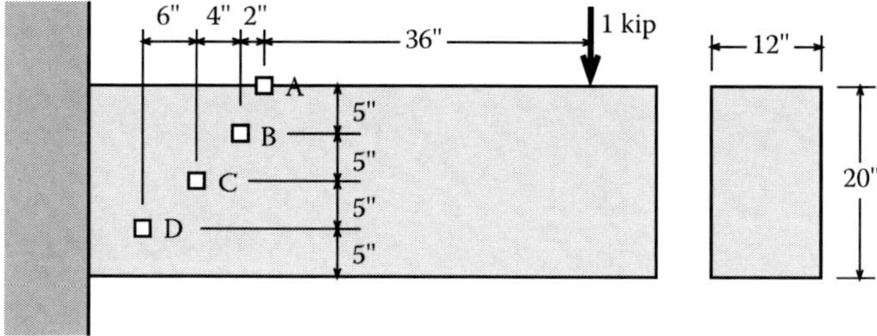

FIGURE 29.6 Cantilever beam.

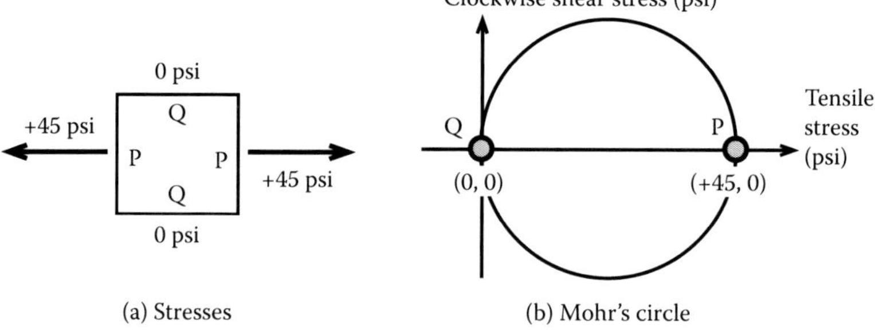

FIGURE 29.7 Stresses at A.

the shear stresses on faces P are zero. Point Q in Figure 29.7b indicates that both the normal and shear stresses on the horizontal faces (Q) are zero.

2. The stress caused by the bending moment at point C is zero. The shear stress, rounding to the nearest integer, is

$$v = \frac{3}{2} \times \frac{V}{bh} = \frac{3}{2} \times \frac{1000 \text{ lbf}}{240 \text{ in.}^2} = 6 \text{ psi}$$

Figure 29.8a shows the stresses acting at C. Figure 29.8b shows the corresponding Mohr's circle. The x-coordinate of point P in Figure 29.8b is 0 because the normal stresses on the vertical faces (P) in Figure 29.8a are 0 psi. The y-coordinate of point P in Figure 29.8b is positive (+6) because the shear stresses on faces P are clockwise. The y-coordinate of point Q in Figure 29.8b is negative (−6) because the shear stresses on faces Q are counterclockwise. If we rotate points P and Q 90 deg in the clockwise direction, as shown in Figure 29.8b, we get points P′ and Q′, representing the principal stresses. Faces P and Q in Figure 29.8a are rotated 90/2 = 45 deg to get faces P′ and Q′ in Figure 29.8c, which shows principal stresses.

3. The stress caused by the bending moment at B, rounded to the nearest integer, is

$$f = \frac{5}{5+5} \times \frac{M}{S} = \frac{1}{2} \times \frac{1000 \text{ lbf} \times 38 \text{ in.}}{800 \text{ in.}^3} = 24 \text{ psi}$$

The shear stress distribution over the depth of a rectangular section is parabolic. The stress caused by the shear force at B is approximately

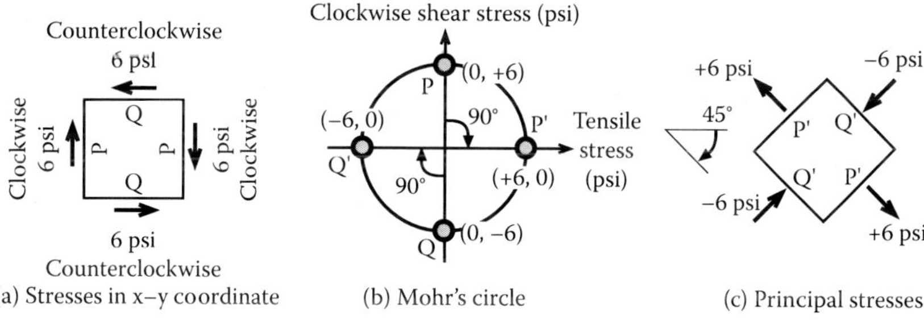

FIGURE 29.8 Stresses at C.

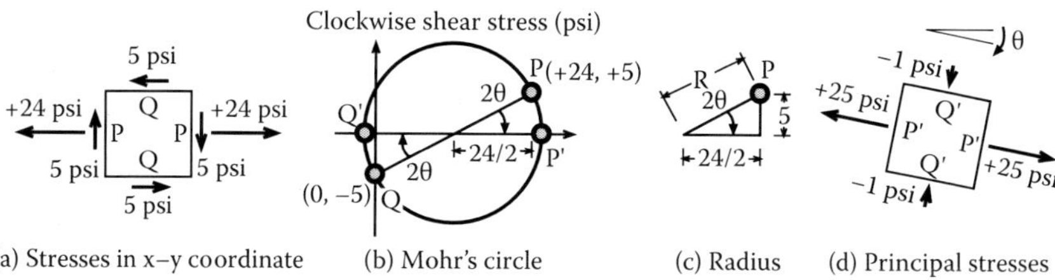

FIGURE 29.9 Stresses at B.

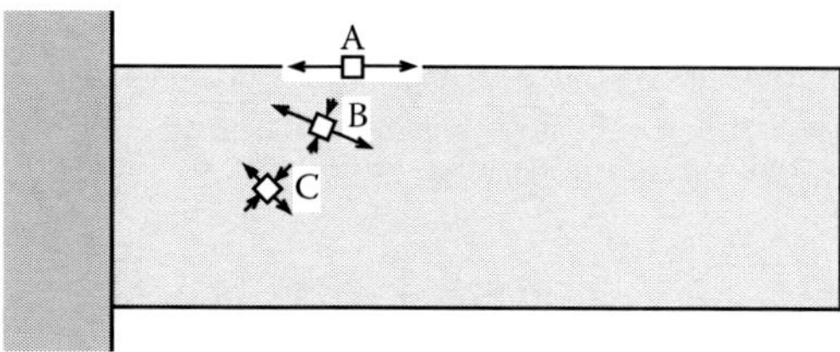

FIGURE 29.10 Principal stresses.

$$v = \left[1 - \left(\frac{1}{2}\right)^2\right] \times \frac{3}{2} \times \frac{V}{bh} = \frac{3}{4} \times \frac{3}{2} \times \frac{1000 \text{ lbf}}{240 \text{ in.}^2} = 5 \text{ psi}$$

Figure 29.9a shows the stresses acting at C. Figure 29.9b shows the corresponding Mohr's circle. The direction of the principal stresses, θ, should satisfy

$$\tan 2\theta = 5 \text{ psi} \div \left(\frac{24 \text{ psi}}{2}\right)$$

Therefore, θ = 11 deg, approximately. The radius of the circle shown in Figure 29.9c is

$$R = \sqrt{\left(\frac{24 \text{ psi}}{2}\right)^2 + (5 \text{ psi})^2} = 13 \text{ psi}$$

The principal stresses are

$$f_1 = \frac{24 \text{ psi}}{2} + 13 \text{ psi} = +25 \text{ psi}$$

and

$$f_2 = \frac{24 \text{ psi}}{2} - 13 \text{ psi} = -1 \text{ psi}$$

The principal stresses are plotted to scale in Figures 29.9d and 29.10.

BARE ESSENTIALS

Longitudinal reinforcement is assumed to provide resistance only in the direction of its axis. While it can compensate for the loss of resistance related to crack formation caused by flexure, it cannot do that efficiently for cracks caused by combined flexure and shear. Transverse reinforcement is used to control these cracks.

Vertical stirrups or hoops are the most common type of transverse reinforcement today.

EXERCISE

Repeat Example 29.1 for point D in Figure 29.6. Check your result using GOYA-MS at the following website: https://nees.org/resources/goyams or http://kitten.ace.nitech.ac.jp/ichilab/mech/en.

Check the "show stress" box, provide the stresses, and rotate the circle using your mouse. Note that each dot on the circle represents the stresses with the same color.

30 Transverse Reinforcement

It is not irrelevant to start this section by quoting Hardy Cross, who exclaimed, when asked about shear in reinforced concrete beams, "Put in stirrups and hang the shear!" The central problem in proportioning reinforced concrete elements to resist combined bending and shear is the determination of how much transverse reinforcement (also called web reinforcement) is required to make certain that the inclined cracks are restrained so that the structural element (be it a column or a beam or a footing) develops its flexural strength and toughness. That is the problem we consider in this chapter. We limit our discussion to monotonically increasing load applied within a short time.

Although much research has been done on the subject, the response of reinforced concrete to shear is still poorly understood. In 1927, after twelve years of work involving a total of 131 tests, F. E. Richart (1927) stated: "The action of reinforcement to resist diagonal tension* is not susceptible of exact analysis."

Richart's statement remains valid, and his work continues to provide us with an excellent, unbiased basis for determining whether and how much web reinforcement is needed.

Consider a beam tested by Richart (Figure 30.1). Beam 224.1 was cast using concrete having a compressive strength of approximately 4000 psi. Its cross section was rectangular, measuring 24 in. in depth and 8 in. in width, with an effective depth, d, of 21 in. Longitudinal reinforcement of the beam was higher than typical, amounting to 2.3%. Each shear span, a, measured 36 in. The beam was reinforced in the transverse (vertical) direction with stirrups, represented as vertical lines in Figure 30.1. In this discussion we shall focus on stirrups perpendicular to the longitudinal axis of the element, usually called vertical stirrups.

To facilitate comparison with other cases, it is convenient to express the amount of transverse reinforcement using the symbol r, the ratio of the total cross-sectional area of *transverse* reinforcement A_w to the product $b \cdot s$, where b is the width of the section* and s is the spacing of the transverse reinforcement (Figure 30.2):

$$r = \frac{A_w}{b \cdot s} \tag{30.1}$$

The beam described in Figure 30.1 had a high transverse reinforcement ratio of r = 1.4%. Figure 30.1 also shows the pattern of cracks observed during testing. In contrast to Figure 10.1, which showed a beam subjected to bending and no shear,

* Diagonal tension refers to the state of unit stress represented in Figure 29.2a.

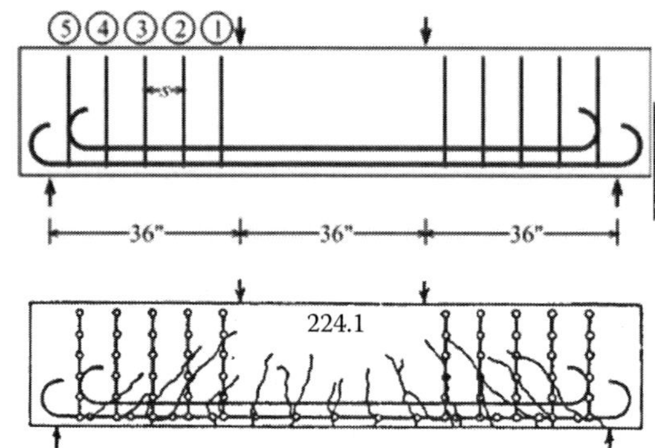

FIGURE 30.1 Beam 224.1 by Richart (1927).

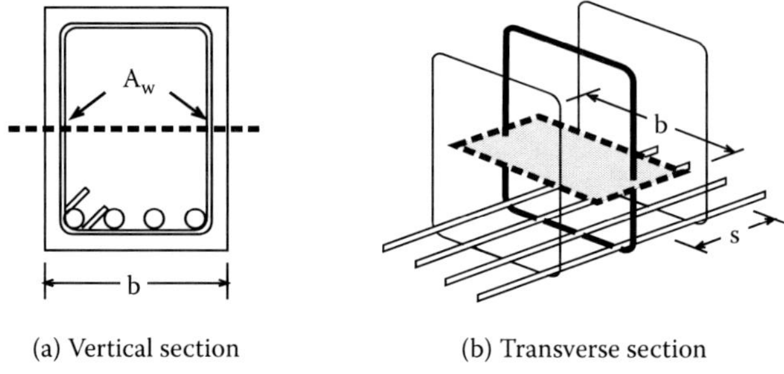

FIGURE 30.2 Definition of transverse reinforcement ratio.

we observe that the combination of shear and moment results in formation of inclined cracks.

The function of the stirrups in relation to shear may be rationalized considering the free-body diagram shown in Figure 30.3. Equilibrium of the beam segment shown requires that $NA_w \cdot f_{sw} = V$, with f_{sw} = stress in transverse reinforcement (N is the number of stirrups crossed by the inclined crack—2 in this case). It is important to remember, while going through this process, that the demand is created by combined bending and shear that leads to tensile stresses inclined to the longitudinal axis of the beam.

Equilibrium of vertical forces can be expressed in terms of unit stress by an undeniably arbitrary definition:

$$\frac{V}{b \cdot d} = N \frac{A_w \cdot f_{sw}}{b \cdot d}$$

where N is the number of stirrups crossed by the crack. Normalizing with respect to $b \cdot d$, as opposed to, say, $b \cdot h$, is mostly a matter of tradition as long as d is close to h. We could not have computed this number a priori from first principles. But from

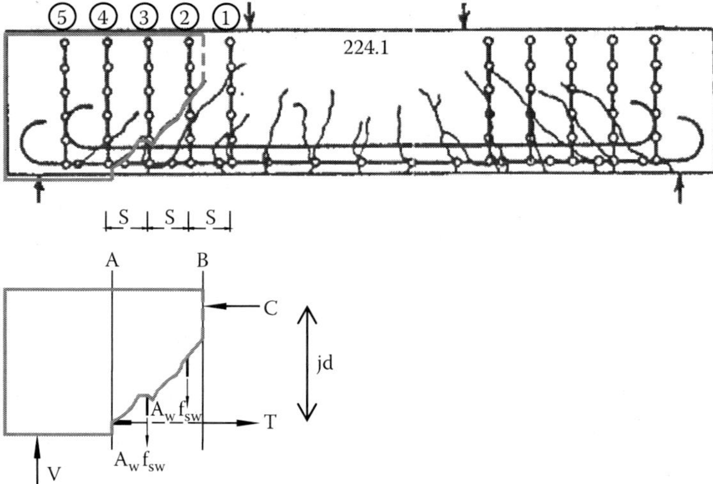

FIGURE 30.3 Free-body diagram.

observation, we know that if s < d/2, inclined cracks that may lead to failure tend to cross, on average, d/s stirrups:

$$\frac{V}{b \cdot d} = \frac{d}{s} \frac{A_w \cdot f_{sw}}{b \cdot d} \tag{30.2}$$

Simplifying and defining $v = \dfrac{V}{bd}$,

$$v = r \cdot t_{sw} \tag{30.3}$$

This expression inspired the tests by Richart in the 1920s. It followed from idcas first formulated in the 1900s by Ritter, who idealized reinforced concrete beams as trusses (in which stirrups act as tension elements, concrete in between inclined cracks acts as compression elements, bottom reinforcement acts as the tension chord, and the concrete in compression acts as the top chord) (Figure 30.4). Theoretical work supported the expression, but the facts did not.

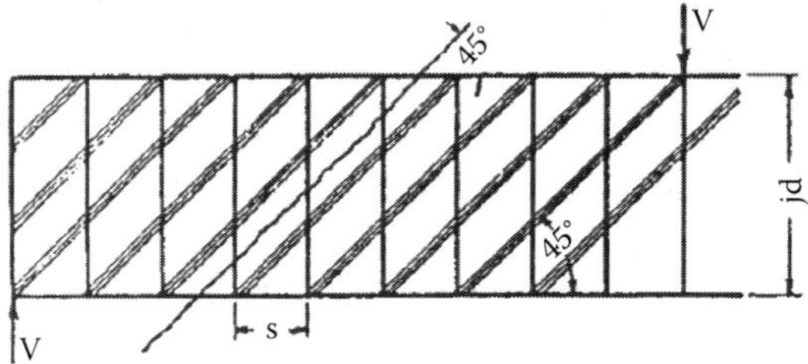

FIGURE 30.4 Truss model by Ritter (1899).

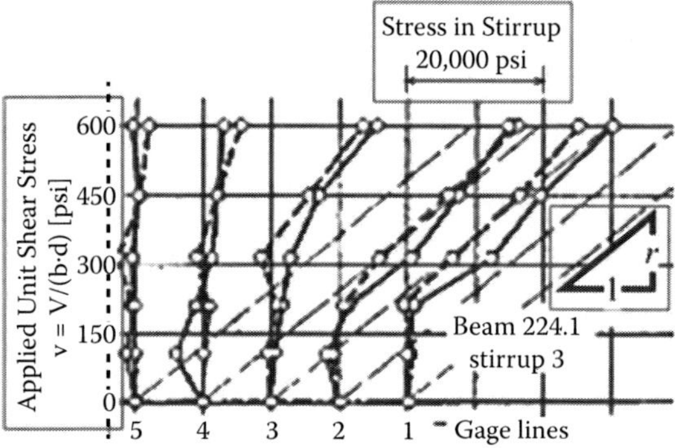

FIGURE 30.5 Stresses in stirrups for the beam in Figure 30.1.

Figure 30.5 shows values of stresses inferred from measurements of deformation in stirrups for the beam in Figure 30.1. The figure shows that the applied stress was not linearly proportional to the stresses in the stirrups measured by Richart, as Equation 30.3 suggests. The measurements showed instead that stresses in stirrups were negligible until the applied unit shear stress exceeded approximately 200 psi (corresponding to the force at an inclined crack crossing the stirrup). Above 200 psi, the stresses in the stirrups did increase at an average rate that approached the expected rate of $v/f_{sw} = r = 0.014$. Richart also observed that the stirrups closer to the support (stirrups 4 and 5) were subjected to much lower stresses than stirrups farther away from the support. This observation indicated that the compression introduced by the support helped to control the width of inclined cracks near the ends of the beam, where the moment and the principal stresses causing inclined cracks were lower. For these reasons, it is generally assumed that if the support applies a compressive force to the bottom of the beam, the critical section for design against shear is located at a distance d from the face of the support.

On the basis of the observations made by Richart and others who have followed, the nominal resistance of a beam to shear is expressed as

$$v_n = C_0\sqrt{f_c'} + r \cdot f_{yw} \tag{30.4}$$

where f_c' is concrete cylinder strength in psi, and f_{yw} is the yield tress of transverse reinforcement.

The first term in this expression is usually referred to as the contribution of concrete to shear strength. It is expressed as a multiple of the square root of the compressive strength of the concrete because shear induces inclined tensile stresses in the concrete, and concrete tensile strength is generally assumed to be proportional to the square root of the compressive strength of concrete, f_c' The second term is the contribution of transverse reinforcement. The terminology has little importance. What is important is to realize that the experimental data have shown that the free-body diagram in Figure 30.3 does not describe all the relevant mechanical actions. Much effort has been invested in quantifying the contributions from mechanisms providing shear resistance different from transverse reinforcement. But for design, it is sufficient to remember that the coefficient C_0 in Equation 30.4, in general, exceeds 1. If the ratio of transverse reinforcement exceeds approximately 0.25%, it is safe to assume $C_0 = 2$.

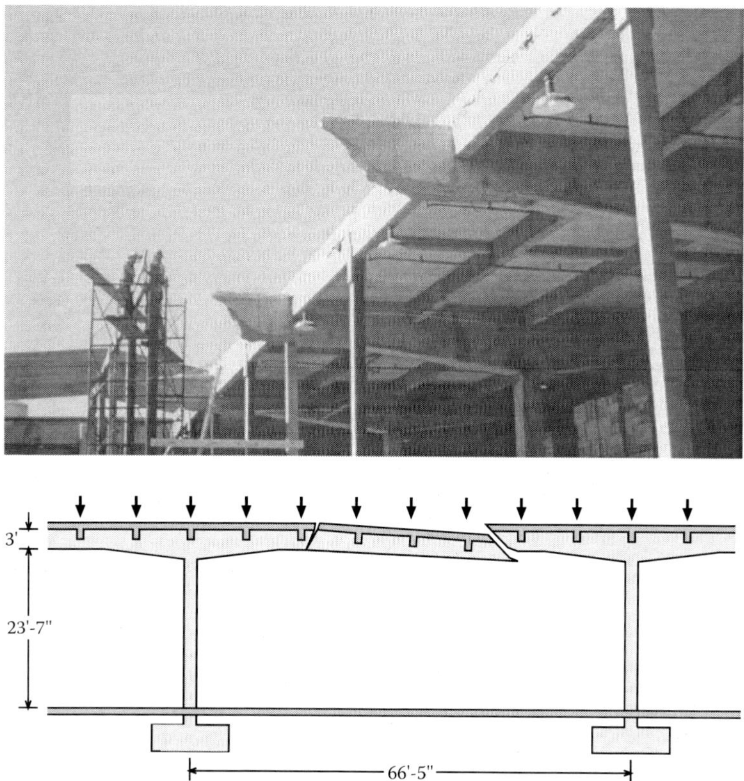

FIGURE 30.6 Failure of beams in a warehouse owned by the U.S. Air Force. (Photo taken by Chester P. Siess.)

In fact, past failures have shown that even if the computed stresses from nominal loads are lower than the resistance attributed to the concrete, one should always use transverse reinforcement. Reinforcement minima are regulated by building codes.

It is also desirable to assume that the nominal shear strength v_n does not exceed an upper bound, which may be on the order of $10\sqrt{f'_c}$ (or as stated by the applicable building code), no matter how much transverse reinforcement is provided.

Shear failure accounts for a large fraction of the unfortunate events in the history of reinforced concrete. Figure 30.6 shows the failure of beams in a warehouse owned by the U.S. Air Force. The failure was attributed to the combined effect of gravity loads and tension induced by shrinkage. The beams had no shear reinforcement. Figure 30.7 shows the photograph of a column in a bridge before and after the 1995 Kobe earthquake. In this case, inertia force caused by earthquake motions caused the shear failure, and the gravity force caused diagonal sliding. The effect of inertia force caused by earthquakes on resistance to shear is beyond the scope of this book. The example from Kobe is included here to alert beginning designers of the importance of shear.

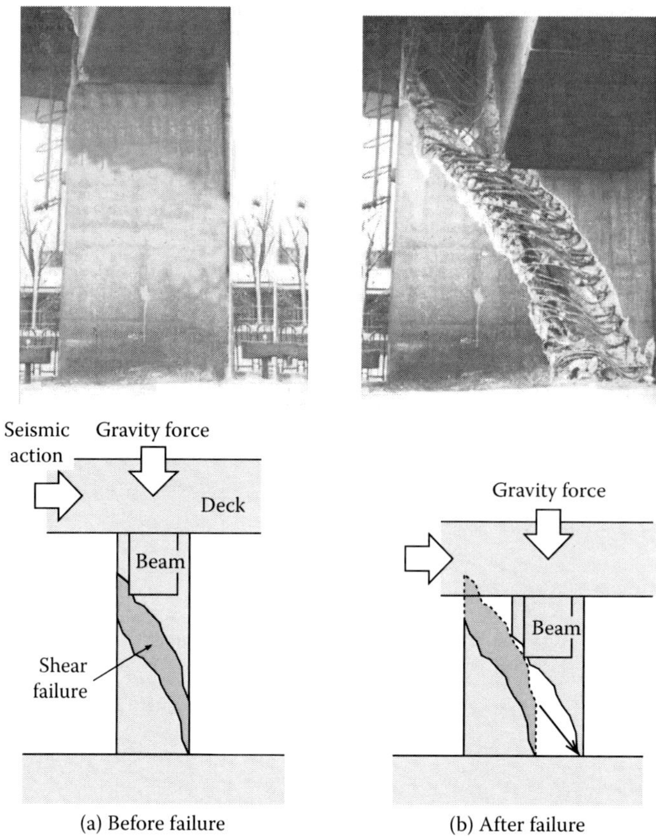

FIGURE 30.7 Reinforced concrete column before and after the Kobe (1995) earthquake.

EXAMPLE 30.1

A reinforced concrete frame is shown in Figure 30.8. The frame is required to resist uniformly distributed loads. Using the governing building code, the factored load to be used for design has been computed to be 8 kip-ft. Multiple numerical analyses of the frame have produced the maximum values of shear shown in Table 30.1. The values reported refer to shear forces at the faces of the columns.

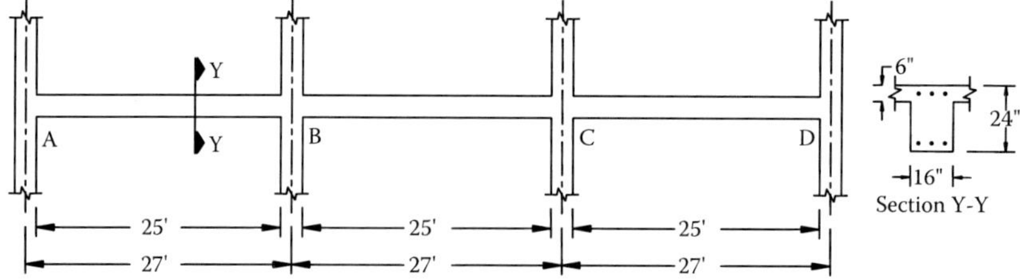

FIGURE 30.8 Reinforced concrete frame to resist uniformly distributed loads.

TABLE 30.1
Maximum Values of Shear at Nodes Indicated in Figure 30.8

Node	A	B_{left}	B_{right}	C_{left}	C_{right}	D
Shear	100	115	100	100	115	100

Note: Unit = kip.

We are asked to select reinforcement to resist shear. Assume $f'_c = 4000$ psi and $f_y = 60$ ksi. Use a strength reduction factor of 0.75.

SOLUTION

First, we examine the values obtained by analysis. We realize that for the exterior spans, the values given do not correspond to a single load arrangement because they do not add up to the total applied load of 8 kip-ft × 25 ft = 200 kip. The values given correspond to maxima from different analyses based on different load distributions throughout the frame with multiple spans. We conclude that we are dealing with an envelope for shear diagrams obtained for different load arrangements. The envelope is shown in Figure 30.9.

We now check the maximum applied shear vs. the maximum resistance. We remember that the critical section is at d from the interior face of the support. So we compute the maximum nominal shear force for design as

$$V_B - w(x) \cdot d = 115\ \text{k} - 8\,\tfrac{\text{kip}}{\text{ft}} \cdot 21\ \text{in.} \cdot \tfrac{1\ \text{ft}}{12\ \text{in.}} = 101\ \text{kip}$$

The maximum nominal unit stress of interest is therefore 101,000 lbf/(16 in. × 21 in.) = 300 psi.

We have been told not to exceed 0.75 × 10 × sqrt(f'_c) = 7.5 × sqrt(4000) psi = 470 psi.

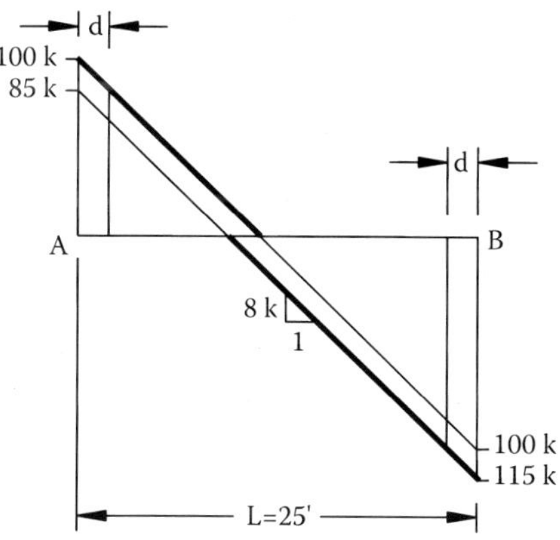

FIGURE 30.9 Envelope for shear diagrams obtained for different load arrangements.

The nominal unit stress from applied loads does not exceed the limit to shear resistance, and so we proceed. We assume that the shear resistance attributable to the concrete is

$$\Phi \cdot 2\sqrt{f_c'} = 0.75 \cdot 2 \cdot \sqrt{4000} \cdot \text{psi} =\sim 95 \text{ psi}$$

The required contribution from the transverse reinforcement needs to exceed (300 – 95) psi = 205 psi. We compute the required ratio, r, of transverse reinforcement to be

$$r = \frac{v_u - \Phi v_c}{\Phi \cdot f_y} = \frac{205 \text{ psi}}{0.75 \cdot 60000 \text{ psi}} = 0.46\%$$

This reinforcement ratio can be provided using #4 stirrups with two legs crossing the potential critical crack. The *maximum* spacing at which these stirrups can be placed is

$$s = \frac{A_w}{b \cdot r} = \frac{2 \times 0.2 \text{ in.}^2}{16 \text{ in.} \cdot 0.0046} = 5.4 \text{ in.}$$

We shall recommend the use of 5 in. spacing to make construction simpler. This spacing is smaller than d/2 (10.5 in.), which indicates to us that it is unlikely that a crack would form without crossing any transverse reinforcement.

We could simply provide #4 stirrups at 5 in. from one another along the entire beam and our task would be complete. But that option may be considered expensive. So we look for a way to increase the spacing of the stirrups. That should be acceptable near the middle of the beam, where the shear demand is low.

The maximum spacing we are recommended to use is (d/2) ~10 in. The reinforcement ratio corresponding to this spacing is

$$r = \frac{A_w}{b \cdot s} = \frac{2 \times 0.2 \text{ in.}^2}{16 \text{ in.} \cdot 10 \text{ in.}} = 0.25\%$$

We could reduce this ratio further by using stirrups made with bars of smaller diameter, but that may lead to errors in construction because using bars with similar but different diameters may cause placement problems. We should also check that this ratio is not below the minimum prescribed by the applicable code (usually ~0.1%).

The question to address next is: Where could the spacing be increased from 5 in. to 10 in.? At the spacing of 10 in. the nominal resistance to shear is

$$\Phi \cdot \left(2 \cdot \sqrt{f_c'} + r \cdot f_y\right) = 0.75 \cdot \left(2 \cdot \sqrt{4000} \text{ psi} + 0.0025 \cdot 60{,}000 \text{ psi}\right) = \sim 200 \text{ psi}$$

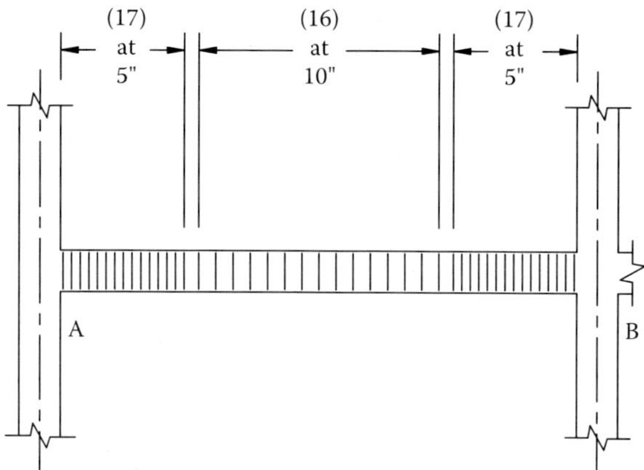

FIGURE 30.10 Resulting reinforcement layout.

In terms of force, that is

$$\Phi v_n \cdot b \cdot d = 200 \text{ psi} \cdot 16 \text{ in.} \cdot 21 = \sim 67 \text{ kip}$$

The computed shear demand reaches this value at

$$(100 - 67) \text{ kip} \div 8 \tfrac{\text{kip}}{\text{ft}} = 4 \text{ ft } 1.5 \text{ in. from the interior face of support A}$$

and

$$(115 - 67) \text{ kip} \div 8 \tfrac{\text{kip}}{\text{ft}} = 6\text{ft from support B}$$

The dimensions we have obtained are not convenient for construction. And design drawings can be difficult to interpret. To keep things simple, we choose to change the spacing of the stirrups to 10 in. in the middle 10 ft of the span and use 5 in. spaces elsewhere. The resulting reinforcement layout is shown in Figure 30.10.

BARE ESSENTIALS

$$v_n = C_0\sqrt{f_c'} + r - f_{yc} \le 10\sqrt{f_c'}$$

If r ≥ 0.25%, it is safe to assume CO = 2.

Spacing of transverse reinforcement should be smaller than d/2.

EXERCISE

The conduit shown in Figures 30.11 and 30.12 carries water and supports a pressurized pipeline. The conduit is made of normal weight concrete. The pipeline is supported at discrete points. At each of these supports the estimated dead load applied on the conduit is 15 kip (unfactored).

Assume

$f'_c = 3000$ psi
$f_y = 60$ ksi
Dead load factor = 1.4
Strength reduction factor for shear = 0.75
Effective depth = 32 in.

Select transverse reinforcement to resist gravity loads. Ignore possible settlement of supports and axial stresses caused by shrinkage.

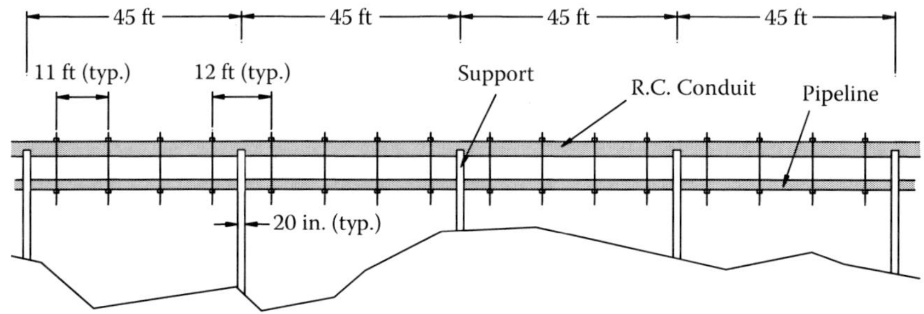

FIGURE 30.11 Elevation of conduit to carry water and support pipeline.

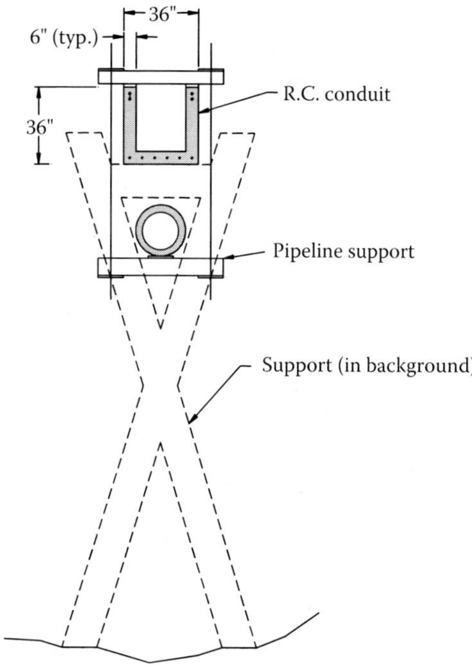

FIGURE 30.12 Cross section of conduit to carry water and support pipeline.

Appendix A: Direct (Three-Point) Solution for a Rectangular Section

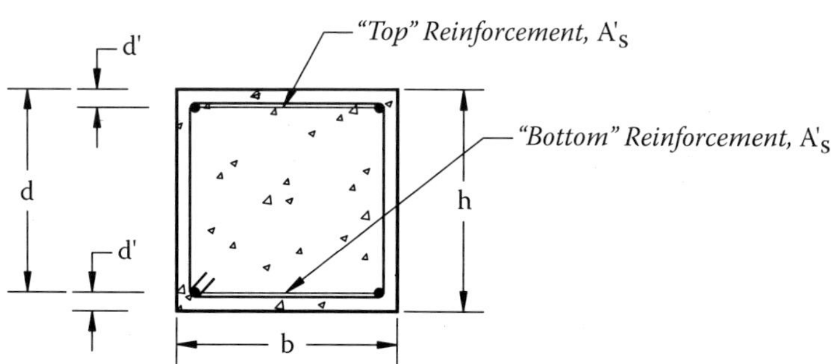

SECTION

Width of Section		$b = 24$ in.
Total Depth of Section		$h = 24$ in.
Depth to Center of Reinforcement		$d' = 3$ in.
Effective Depth	$d = h - d'$	$d = 21$ in.
Top Reinforcement		$A'_s = 3$ in.^{2}
Bottom Reinforcement		$A_s = 3$ in.^{2}
Yield Stress		$f_y = 60$ ksi
Young's Modulus for Steel Reinforcement		$E_s = 29000$ ksi
Yield Strain	$\varepsilon_y = \dfrac{f_y}{E_s}$	$\varepsilon_y = 2.1 \times 10$
Concrete Compressive Strength		$f'_c = 4000$ psi
Young's Modulus for Concrete		
	$E_c = 57000\sqrt{f'_c}$	$E_c = 3.6 \times 10^6$ psi
Limiting Strain in Compression for Concrete		$\varepsilon_{cu} = 0.003$

PLASTIC CENTROID

For determining the moment capacity about an axis, the plastic centroid is defined by the distance perpendicular to the axis from either extremity of the section. In the case of a rectangular section, we shall define it with respect to the top of the section. The plastic centroid is the point (or the axis) about which the fully developed stresses in the material generate zero moment.

The fully developed stress for the concrete is assumed to be 85% of the cylinder strength. The fully developed stress for the reinforcement is taken to be its yield stress.

It is to be understood that the plastic centroid is simply a definition of convenience to locate a position in the section about which to state the resisting moment. As long as the engineer using the calculated resisting moment is aware of the location about which the moments have been taken, the plastic moment can be at any point.

Take moments of fully developed sections about top of section

Concrete is assumed to have the strength $f''_c = 0.85\ f'_c$

$f''_c = 3.4 \times 10^3$ psi

Moment generated by concrete about top of section

$$M_c = \frac{b \cdot h^2}{2} \cdot f''_c \qquad M_c = 2 \times 10^3 \text{ kip-ft}$$

Moment generated by reinforcement about top of section

$$M_r = (A'_s \cdot d' + A_s \cdot d) \cdot (f_y - f''_c) \qquad M_r = 340 \text{ kip-ft}$$

Depth to plastic centroid from top

$$d_p = \frac{M_c + M_r}{b \cdot h \cdot f''_c + (A_s + A'_s) \cdot (f_y - f''_c)} \qquad d_p = 12 \text{ in.}$$

Because the section treated is symmetrical, we could have guessed at d_p. It should be at mid-height. The routine above can be used to obtain a perspective of the effect of unsymmetrical arrangement of reinforcement on location of the plastic centroid.

Now we start a set of three simple calculations to determine the moment and axial load for three bounding states:

STATE 1: AXIAL LOAD

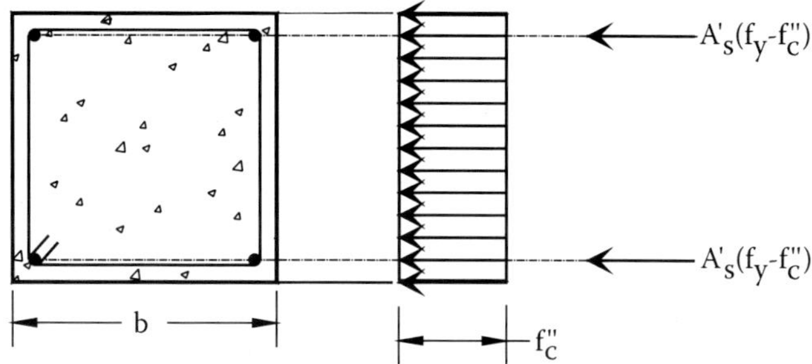

$$P_o = b \cdot h \cdot f''_c + (A'_s + A_s) \cdot (f_y - f''_c)$$

$$P_o = 2.30 \times 10^3 \text{ kip}$$

STATE 2: "BALANCED" CONDITION

Defined by the simultaneous occurrences of limiting compressive strain at the extreme fiber in compression and yield strain at the bottom tensile reinforcement.

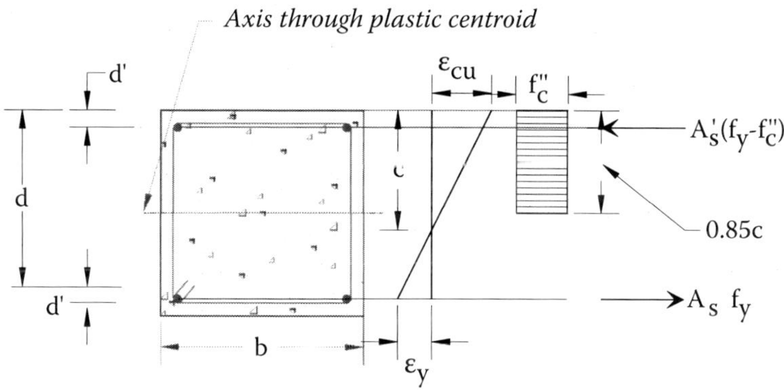

Depth to Neutral Axis

$$c = \frac{\varepsilon_{cu}}{\varepsilon_y + \varepsilon_{cu}} \cdot d \qquad a = 12.4 \text{ in.}$$

Strain in Top Bar

$$\varepsilon'_s = \frac{\varepsilon_{cu} \cdot (c - d')}{c} \qquad \varepsilon'_s = 0.0023$$

Stress in Top Bars

$$f'_s = \begin{vmatrix} \varepsilon'_s \cdot E_s \text{ if } \varepsilon'_s \le \varepsilon_y \\ f_y \text{ otherwise} \end{vmatrix} \qquad f'_s = 60 \cdot \text{ksi}$$

Stress in Bottom Bars

by definition $\qquad f_y = 60 \text{ ksi}$

Total Force in Concrete

$$C_c = b \cdot 0.85c \cdot f''_c \qquad C_c = 862 \text{ kip}$$

Resisting Moment about Plastic Centroid

$$M_c = C_c \cdot \left(\frac{h}{2} - \frac{0.85 \cdot c}{2}\right) \qquad M_c = 483 \text{ kip-ft}$$

$$M_s = A_s \cdot f_y \cdot (d - d') \qquad M_s = 270 \text{ kip-ft}$$

$$M_{bal} = M_c + M_s \qquad M_{bal} = 753 \text{ kip-ft}$$

Axial Force Resistance

$$P_{bal} = C_c + A'_s \cdot f'_s - As \cdot f_y$$

$$P_{bal} = 862 \text{ kip}$$

STATE 3: AXIAL FORCE ~0

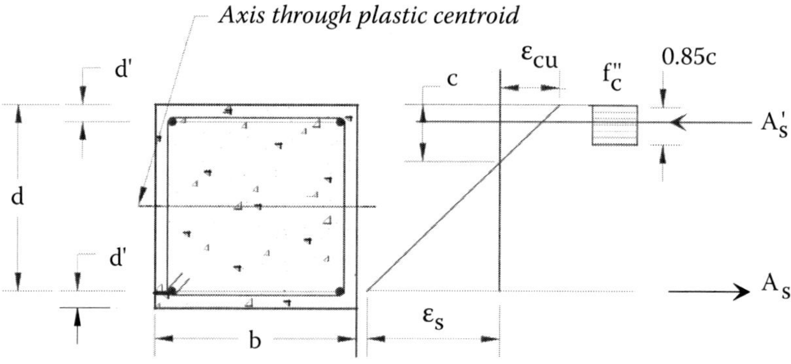

Determine strain distribution for P = 0 for a depth to neutral axis of c.

Strain in Top Reinforcement

$$\varepsilon_s'(c) = \frac{\varepsilon_{cu}}{c} \cdot (c - d')$$

Stress in Top Reinforcement

$$f_s'(c) = \begin{vmatrix} f_y \text{ if } \varepsilon_s'(c) \geq \varepsilon_y \\ \varepsilon_s'(c) \cdot E_s \quad \text{otherwise} \end{vmatrix}$$

Force in Concrete

$$C_c(c) = b \cdot 0.85\ c \cdot f_c''$$

Net Axial Force

Ignoring the correction for area of concrete occupied by the top reinforcement

$$\text{NetForce}\ (c) = A_s' \cdot f_s'(c) + C_c(c) - A_s \cdot f_y$$

Assume

$c = 2.83$ in.

NetForce (c) = 0 kip

Moment about Plastic Centroid

$$M_3(c) = A_s \cdot f_y \cdot \left(d - \frac{h}{2}\right) + A_s' \cdot \left(\frac{h}{2} - d'\right) \cdot f_s'(c) + C_c(c) \cdot \left(\frac{h}{2} - 0.85 \cdot \frac{c}{2}\right)$$

$M_3(c) = 299$ kip-ft

Note that for this case M_3 could be approximated simply as

$M_{3a} = 0.9\ A_s \cdot f_y \cdot (d)$ $M_{3a} =$ 283 kip-ft

SUMMARY

Axial force

$$P = \begin{pmatrix} P_o \\ P_{bal} \\ 0 \text{ kip} \end{pmatrix} \qquad P = \begin{pmatrix} 2298 \\ 862 \\ 0 \end{pmatrix} \cdot \text{kip}$$

Bending moment (about plastic centroid)

$$M = \begin{pmatrix} 0 \\ M_{bal} \\ M_3(c) \end{pmatrix} \qquad M = \begin{pmatrix} 0 \\ 753 \\ 299 \end{pmatrix} \cdot \text{kip} \cdot \text{ft}$$

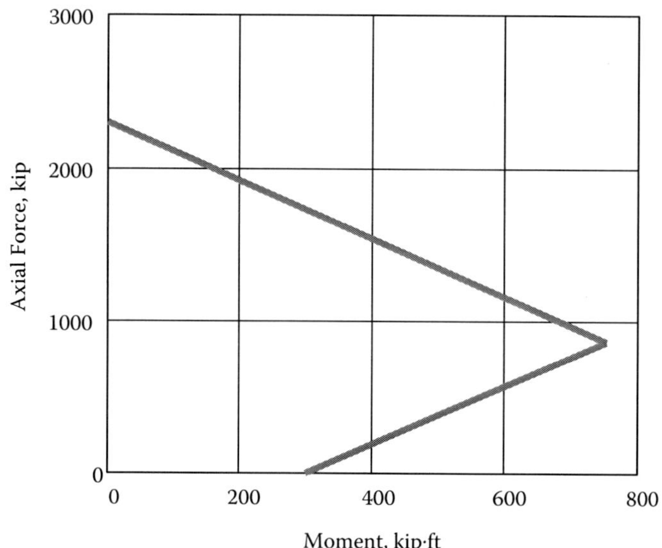

Appendix B: Iterative Solution for a Rectangular Section

INPUT

Width of Section	$b = 24$ in.
Total Depth of Section	$h = 24$ in.
Depth to Top Reinforcement	$d' = 3$ in.
Effective Depth	$d = 21$ in.
Top Reinforcement	$A'_s = 3$ in.^{2}
Bottom Reinforcement	$A_s = 3$ in.^{2}
Column Reinforcement Ratio $\rho_c = \frac{A'_s + A_s}{b \cdot h}$	$\rho_c = 1\%$
Yield Stress	$f_y = 60$ ksi
Young's Modulus for Steel	$E_s = 29000$ ksi
Yield Strain	$\varepsilon_y = 0.002$
Concrete Compressive Strength	$f'_c = 4000$ psi
Young's Modulus for Concrete $E_c = 57000\sqrt{f'_c}$	$E_c = 3.6 \times 10^6$ psi
Limiting Strain for Concrete	$\varepsilon_{cu} = 0.003$

PLASTIC CENTROID

Take moments of fully developed stresses about the top of section:

Concrete is assumed to have the strength $f''_c = 0.85\ f'_c$

Moment generated by concrete

$$M_1 = \frac{b \cdot h^2}{2} \cdot f_c'' \qquad M_1 = 2 \times 10^3 \text{ kip-ft}$$

Moment generated by reinforcement

$$M_2 = (A_s' \cdot d' + A_s \cdot d) \cdot (f_y - f_c'') \qquad M_2 = 340 \text{ kip-ft}$$

Depth to plastic centroid from top

$$d_p = \frac{M_1 + M_2}{b \cdot h \cdot f_c'' + (A_s + A_s') \cdot (f_y - f_c'')} \qquad d_p = 12 \text{ in.}$$

Because the section treated is symmetrical about a horizontal axis, we could have guessed at d_p.

CAPACITY FOR AXIAL LOAD, P_O

The axial-load capacity is simply the sum of the loads that can be resisted at the assumed limiting stresses of the reinforcement and the concrete.

$$P_0 = b \cdot h \cdot f_c'' + (A_s' + A_s) \cdot (f_y - f_c'') \qquad P_0 = 2.3 \times 10^3 \text{ kip}$$

COMBINATIONS OF AXIAL FORCE AND BENDING MOMENT

We initiate an iterative procedure to determine the moment and axial load at various values of tensile strain at the level of the "bottom" reinforcement, ε_s.

We start with

$Net\_Force_0 = 1\ kip$

$$
IA = \left|
\begin{array}{l}
\text{while } Net\_Force_i > 0 \\
\quad \left|
\begin{array}{l}
i \leftarrow i+1 \\
\varepsilon_{si} \leftarrow i \cdot \dfrac{\varepsilon_y}{10} - \dfrac{\varepsilon_y}{10} \\
c_i \leftarrow \dfrac{\varepsilon_{cu}}{\varepsilon_{si} + \varepsilon_{cu}} \cdot d \\
\varepsilon'_{si} \leftarrow \dfrac{c_i - d'}{c_i} \cdot \varepsilon_{cu} \\
f'_{si} \leftarrow \left|
\begin{array}{ll}
(\varepsilon'_{si}) \cdot E_s & \text{if } |\varepsilon'_{si}| \le \varepsilon_y \\
f_y \cdot \dfrac{\varepsilon'_{si}}{|\varepsilon'_{si}|} & \text{otherwise}
\end{array}\right. \\
f_{si} \leftarrow \left|
\begin{array}{ll}
E_s \cdot \varepsilon_{si} & \text{if } E_s \cdot \varepsilon_{si} \le f_y \\
f_y & \text{otherwise}
\end{array}\right. \\
Reinf\_Force_i \leftarrow A'_s \cdot f'_{si} - A_s \cdot f_{si} \\
Conc\_Force_i \leftarrow 0.85 \cdot b \cdot c_i \cdot f''_c \\
Net\_Force_i \leftarrow Reinf\_Force_i + Conc\_Force_i \\
Reinf\_Mom_i \leftarrow A'_s \cdot f'_{si} \cdot (d_p - d') + A_s \cdot f_{si} \cdot (d - d_p) \\
Conc\_Mom_i \leftarrow Conc\_Force_i \cdot \left(d_p - \dfrac{0.85\ c_i}{2} \right) \\
Total\_Mom_i \leftarrow Reinf\_Mom_i + Conc\_Mom_i \\
Store_{i,0} \leftarrow \dfrac{Total\_Mom_i}{kip\text{-}ft} \\
Store_{i,1} \leftarrow \dfrac{Net\_Force_i}{kip} \\
Store_{i,2} \leftarrow i
\end{array}\right. \\
Store
\end{array}\right.
$$

a

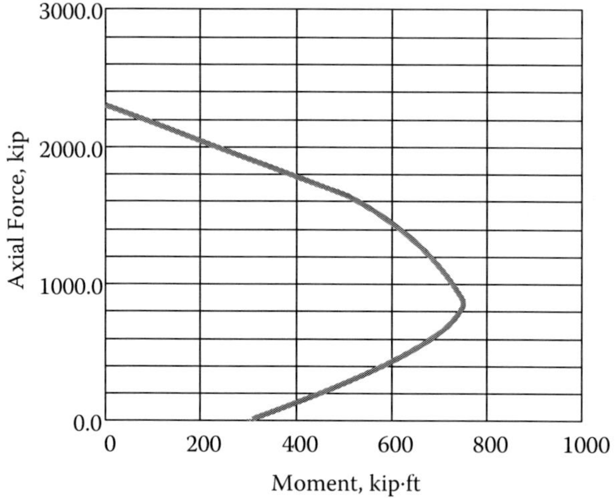

We compare the current "detailed" solution with the previous (three-point) simple one:

$$P = \begin{pmatrix} 2298 \\ 862 \\ 0 \end{pmatrix} \qquad M = \begin{pmatrix} 0 \\ 753 \\ 299 \end{pmatrix}$$

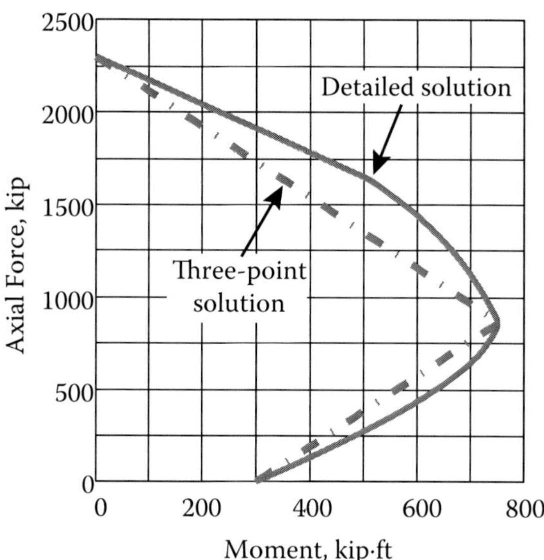

References

Abrams, D. A. (1913). *Tests of Bond between Concrete and Steel.* University of Illinois Engineering Experiment Station Bulletin 71. University of Illinois at Urbana–Champaign, Urbana–Champaign, Illinois.

Abrams, D. A. (1918). *Design of Concrete Mixtures.* Structural Materials Research Laboratory, Lewis Institute, Chicago, Illinois.

ACI Committee 105. (1933). Reinforced Concrete Column Investigation, Final Report. *Proceedings of ACI*, 29, p. 275.

Broms, B. B. (1965). Crack Width and Crack Spacing in Reinforced Concrete Members. *ACI Journal, Proceedings*, 62, 10, 1237–1255.

Carneiro, F. (1947). Une Nouvelle Methode d'Essai pour Determiner la Resistance a La Traction du Beton. *Reunion des Laboratoires.* Paris, France.

CEB-FIP. (1973). *Structural Effects of Time-Dependent Behavior of Concrete.* Brookfield Publishing Company, Philadelphia, Pennsylvania.

Corley, W. G., and Sozen, M. A. (1966). Time Dependent Deflections of Reinforced Concrete Beams. *ACI Journal Proceedings*, 63, 3.

Cross, H. (1930). *The Column Analogy.* University of Illinois Engineering Experiment Station Bulletin 215. University of Illinois at Urbana–Champaign, Champaign–Urbana, Illinois.

Cross, H., and Morgan, N. D. (1932). *Continuous Frames of Reinforced Concrete.* Wiley, New York.

deVries, R. A. (1989). Lap-Splice Strength of Plain and Epoxy-Coated Reinforcement, *Structural Engineering.* School of Civil Engineering, University of California at Berkeley, Berkeley, California.

Kluge, R. W., and Tuma, E. C. (1945). Lapped Bar Splices in Concrete Beams. *Journal of the American Concrete Institute*, 42, 9, 13–34.

Mains, R. M. (1951). Measurement of the Distribution of Tensile and Bond Stresses along Reinforcing Bars. *Journal of ACI*, 23, 3, 225–252.

Moersch, E. (1902). Der Betoneisenbau, seine Anwendung und Theorie. Wayss & Freytag, Stuttgart.

Rehm, G. (1961). *Über die Grundlagen des Verbundes zwischen Stahl und Beton.* Deutscher Auschuss für Stahlbeton, Heft 138, Berlin, Germany.

Richart, F. E. (1927). *An Investigation of Web Stresses in Reinforced Concrete Beams.* University of Illinois Engineering Experiment Station Bulletin 166, pp. 1–105. University of Illinois at Urbana–Champaign, Champaign–Urbana, Illinois

Richart, F.E., and Brown, R.L. (1934). An Investigation of Reinforced Concrete Columns. *University of Illinois Engineering Experiment Station Bulletin* 267, pp. 1–91. University of Illinois at Urbana–Champaign, Champaign–Urbana, Illinois.

Richart, F. E., Brandtzaeg, A., and Brown, R. L. (1929). The Failure of Plain and Spirally Bound Concrete in Compression, *University of Illinois Engineering Experiment Station Bulletin* 190. University of Illinois at Urbana–Champaign, Champaign–Urbana, Illinois.

Richter, B. (2012). A New Perspective on the Tensile Strength of Lap Splices in Reinforced Concrete Members. MS thesis. Purdue University, West Lafayette, Indiana.

Ritter, W. (1899). Die Bauweise Hennebique. *Schweizerische Bauzeitung*, 17, 41–61.

Ross, A. D. (1958). Creep of Concrete under Variable Stress, *Journal of the American Concrete Institute*, 29, 9, 739–758.

Sozen, M. A., and Siess, C. P. (1963). Investigation of Multiple-Panel Reinforced Concrete Floor Slabs; Design Methods—Their Evolution and Comparison. *ACI Journal, Proceedings*, 60, 8, 999–1028.

Sozen, M. A., and Moehle, J. P. (1990). Development and Lap-Splice Lengths for Deformed Reinforcing Bars. In *Concrete, A Report to the Portland Cement Association, Skokie, IL, and the Concrete Reinforcing Steel Institute*, Schaumburg, Illinois.

Talbot, A. N. (1904). *Tests of Reinforced Concrete Beams*. University of Illinois Bulletin 1, p. 25. University of Illinois at Urbana–Champaign, Champaign–Urbana, Illinois

Talbot, A. N. (1907). *Tests of Reinforced Concrete Beams*. University of Illinois Bulletin 14, p. 21. University of Illinois at Urbana–Champaign, Champaign–Urbana, Illinois

Treece, R.A. and Jirsa, J.O. (1989). Bond Strength of Epoxy Coated Reinforcing Bars. *American Concrete Institute Materials Journal*, 86, 2, March-April.

Index

D

T

U

V

W

Y